Prof. Dr. Dr. hon. causa Janos Ladik
Dr. Wolfgang Förner

Universität Erlangen
Institut für Theoretische Chemie
Egerlandstraße 3
91058 Erlangen

ISBN 3-540-57962-1 Springer-Verlag Berlin Heidelberg New York
ISBN 0-387-57962-1 Springer-Verlag New York Berlin Heidelberg

CIP-data applied for

This work is subject to copyright. All rights are reserved, whether the whole or part of the material is concerned, specifically the rights of translation, reprinting, reuse of illustrations, recitation, broadcasting, reproduction on microfilm or in other ways, and storage in data banks. Duplication of this publication or parts thereof is permitted only under the provisions of the German Copyright Law of September 9, 1965, in its current version, and permission for use must always be obtained from Springer-Verlag. Violations are liable for prosecution act under German Copyright Law.

© Springer-Verlag Berlin Heidelberg 1994
Printed in Germany

The use of general descriptive names, registered names, trademarks, etc. in this publication does not imply, even in the absence of a specific statement, that such names are exempt from the relevant protective laws and regulations and therefore free for general use.

The publisher cannot assume any legal responsibility for given data, especially as far as directions for the use and the handling of chemical are concerned This information can be obtained from the instructions on safe laboratory practice and from the manufacturers of chemicals and laboratory equipment.

Typesetting: Camera-ready by authors
SPIN:10123876 51/3020-5 4 3 2 1 0 - Printed on acid -free paper

J. Ladik · W. Förner

The Beginnings of Cancer in the Cell

An Interdisciplinary Approach

With 47 Figures

Springer-Verlag
Berlin Heidelberg New York
London Paris Tokyo
Hong Kong Barcelona Budapest

Contents

Preface

With the great advances in biology in the last decades we have become rather close to understanding - at least on the biochemical level - the onset of cancer in the biological cell. The goal of this book is to describe our present knowledge about the start of the cancerous change in the single cell in a qualitative form understandable to biologists, chemists and possibly to educated laymen. It should be emphasized already here that the onset of cancer in a single cell is a complex but purely scientific problem and it becomes a medical problem only, when due to the accumulation of cancerous cells, a tumor starts to develop in an organism.

In the last decade more than 100 cancer causing genes (so-called oncogenes) have been discovered in human tumors. This has proven the more than 15 year old suspicion that in the final analysis we have the cause of cancer in our genetic material. On the other hand these oncogenes are usually dormant for most of the time and need external factors (such as cancer triggering chemicals, the so-called chemical carcinogens, or various types of radiation) to become active.

The main part of the book contains the description how these external factors activate the oncogenes. This discussion proceeds, starting from the well known biochemical facts to the most probable physical and physico-chemical ways through which these external factors can interfere with the normal interaction of DNA (the genetic material) and proteins. Namely the regulation of genes (including oncogenes) is determined by DNA-protein interactions and the disturbance of these interactions changes also the genetic regulation. As theoreticians we think not only on local, mostly chemical effects at their point of attack of the external cancer starting factors, but also on their long range effects along the DNA chain. To understand these long range effects one has to consider the giant molecules which play a key role in the cell, like DNA and proteins, as being complicated solids. The scientific treatment of the problem needs the methods of theoretical solid state physics, but in this book only the results will be described in a non-mathematical form which can be easily understood by non-theoreticians.

In the subsequent chapters of the book it will be shown (again in a non-mathematical qualitative way) how the activation of oncogenes can disturb the self-regulation of a biological cell to such an extent that the overall state of the cell starts to change (which corresponds to the onset of cancer in a single cell). One of the later chapters is devoted to the experimentally found connection in higher organisms between the start of cancer in the cells and changes in the brain. This puts the problem into a new and more general context.

The final part of the book strongly advocates that to understand still more deeply the start of the cancerous change in a single cell, a very basic interdisciplinary research including the whole spectrum of sciences (mathematics, physics, physical chemistry, chemistry, and biology) is needed. It is highly probable that on the basis of the results of such a (both experimental and theoretical) research project new and more effective methods for the prevention of cancer could be worked out in close cooperation with doctors of medicine.

The authors would like to express their deep gratitude to the late Professor Albert Szent-Györgyi, N.L., whose inspiration has contributed very much to the formation of the ideas decribed in this book. They are further very grateful to Professor N. Fiebiger, former President of the Friedrich-Alexander University Erlangen-Nürnberg whose continous interest and support was a permanent source of strength.

They are very much indebted to the late Professor K. Laki, to Professors M. Blohmke, I. B. Weinstein, D. Grunberger and Dr. A. von Metzler from whom they have learned very much about the newest molecular biological developments in the problem of the onset of cancer and about its connection with the brain.

From the theoretical side they have profited very much from the continuous cooperation and discussions with Professors J. Čížek, E. Clementi, T. C. Collins, G. Del Re, W. Forbes, T. A. Hoffmann, M. Lax, F. Martino, P. G. Mezey, and M. Zaider. They are very much indebted to all their present and former colleagues in the Institute in Erlangen, especially to Professor P. Otto, Dr. S. Suhai and Professor M. Seel. Further, one of us (W.F.) wants to thank Dipl. Chem. Reinhard Knab for his help with the preparation of the final forms of the chemical formulas, necessary for his writing of Chapters 1-3 and Dr. Detlef Hofmann for the supply of part of the literature on radiation damage.

Special thanks are due to Dr. F. Salisbury, J. D. LL. D. (Hon.), President and Chief Executive Officer and to Mrs. T. Salisbury, Vice-president of the National Foundation of Cancer Research. Without the continuous and substantial financial help of NFCR, the major part of the theoretical investigations described in this book could not have been performed. They are further very much indebted to the Siemens AG Company, as well as to IBM Germany and IBM USA (Kingston, N.Y.) who have all provided substantial amounts of free computer time for the large scale calculations connected with the cancer problem. They are also very grateful to the Alexander von Humboldt Foundation, to the "Deutsche Forschungsgemeinschaft" and to the "Deutscher Akademischer Austauschdienst" who provided Visiting Professorships and other Fellowships for Scientists who have taken part in the theoretical work connected with the start of cancer.

The authors are very grateful to Dr. E. Häusser, President of the German Patent Office and of the German Institute for Innovation, to Dipl. Ing. U. Poppe, for many years director of the German Office of Innovation and to the late Professor E. Krowowski who did their best to find financial support for the planned and necessary experiments in Germany.

The authors would like to express their special thanks to Dr. R. W. Stumpe, Springer Publishing Co., not only for publishing this book in an appropriate form, but also for strongly encouraging us to write it.

Last but not least, we should like to thank very much Mrs. Eva Ladik, the wife of J. L., for her patience during the preparation of this manuscript and for drawing all the figures in such a professional way.

János Ladik
Wolfgang Förner
Erlangen, December 1993

Introduction (The History of Cancer)

Cancer is older than mankind. S. U. Willinston, in the early 1920s, found a bone tumor on the skeleton of a dinosaur in Wyoming [1]. Since dinosaurs lived a long time before humans - in the Triassic, Cretaceous and Jurassic periods, while the first remains of humans can only be dated back to the end of the Tertiary period - cancer must have existed in the biosphere of the earth well before mankind evolved.

D. Brothwell [2] found a bone tumor in a 1500000 year old human jaw in the British Museum of London, while C. W. Goodman and G. M. Moran discovered the signs of a malignant bone tumor in the septum hole in the skeleton of a man from the Neolithic period in Dorset (England) [3]. As this shows, even stone age men suffered from cancer.

In Egypt, though the average life expectancy was less than 30 years (the probability of having cancer increases with the fourth power of age, that is in a population group twice as old the occurence of cancer is $2^4=16$ times higher), a bone cancer was found in an Egyptian mummy from the fifth dynasty (3160-2920 B.C.) [4]. In 1862, in a tomb near Thebes, a papyrus was found from 1500 B.C. which describes different diseases and methods for their medical treatment. It also mentions tumors in human tissues which can occur in all parts of the body [5].

Cancer was known also in Mesopotamia before Christ [6] and the Bible [7] mentions that a prophet said that King Joram would get a disease. This came about and the king became sick with an incurable disease of his intestines. According to Preuss [8], this disease was very probably cancer of the intestines. Other ancient Jewish sources also mention the occurrence of different types of malignant tumors. According to Buhac [9], Herod the Great very probably died from a pancreas carcinoma. It should be pointed out also that cancer was known both in ancient China (mentioned in about 660 B.C.) and India [10].

Cancer was mentioned in Greece even before the birth of Christ by Herodotos [11]. According to him, a Greek doctor cured the wife of the emperor Darius from a cancer of the breast. The great medical doctor Hippocrates, who worked and established his school on the island of Kos, described in his work [12] the causes and treatment of different types of cancer (mainly breast and stomach cancers).

The knowledge of Greek medicine was brought to Rome, and with it to the whole Roman Empire, by several Greek doctors among whom perhaps Asklepiades (who arrived in Rome in 91 B.C.) is the most well known. Numerous cases of cancer are reported in the works of different Roman authors. For instance Cornelius Nepos wrote that a friend of his developed a cancer of the intestines which the man tried to cure through fasting. In the time of Ovids (43 B.C. - 17 A.D.), as he mentioned in his Metamorphoses, cancer was a frequently occurring incurable disease. For instance the teacher of the emperor Nero in his youth most probably died because of a cancer of the throat [13]. The death of emperor Vespasianus was caused most probably by a brain tumor [8]. Galen (born in 130 A.D. in Pergamon) was the most important Roman doctor whose medical authority dominated the Middle Ages. Galen believed that cancer starts in the gall bladder and occurs most frequently in the breasts of women. He recommended an operation in the early stages of cancer, and in the later stages, its burning out with a very hot iron [14]. He recommended the medicine Theriak against cancer. Theriak originated from the Parthian king Mithridates VI and his doctor and contained many components including the poison of snakes. This mixture was produced throughout the Middle Ages and it was also available in the early part of this century in some pharmacies in Dalmatia. Despite all the efforts of their doctors also many prominent Romans died from cancer, the Emperor Diocletianus among them.

There are also many descriptions of cancer cases in works which come from the Byzantine Empire from which the books of Oreibasios (written following the order of emperor Julianus the "Apostata") may be mentioned [15]. He also thought cancer starts in the gall bladder and can best be cured at an early stage by an operation.

After the fall of the Western Roman Empire and the decline of the Byzantine Empire, the focus of culture including medicine shifted to the Arab countries. The Arabs took over the Greek-Roman heritage of medicine and made their own important contributions to it. Quite a few Arab doctors wrote about cancer. Among them the great Avicenna wrote in his book Kânûn [16] that melancholy (depression) can cause cancer and made quite a few suggestions for the cure of cancerous tumors in their early stage [16]. Besides their own contributions, the main role of the Arabs was the transmission of the knowledge of the ancient world to the Europe of the late Middle Ages.

There are a large number of books about cancer and its treatment from the Middle Ages, from the Renaissance and Baroque times, as well as from the more recent past (18th and 19th century). The discussion of this literature falls outside the scope of this book, but the interested reader can refer to the book by Körbler [5] which contains a large number of original references.

We make only two exceptions here by mentioning the work of Paracelsus and the discovery of the first chemical carcinogens (cancer-causing chemical substances) in Britain in the 19th century. Paracelsus who was born in the middle of the 15th century (1453) was a real polymath (scholar, physician, alchemist). He wrote among many other books on medicine alone 46 books [17]. He pointed out very strongly that medicine is an empirical science which collects the relevant experiences of many people to cure the different diseases. At the same time, a good doctor has to possess a very good intuition to come to the right diagnosis. He believed, in contrast to his ancestors, that cancer has a chemical origin and he trusted that it can be cured also by other chemicals which he tried to discover on the basis of ordinary people's knowledge. He was strongly against the operation of tumors.

England in the 19th century experienced great developments in medicine including cancer research. Paris discovered as early as 1822 that arsenic can cause cancer [18]. They discovered also that after the introduction of mineral oil in the wool industry between 1850 and 1870 the number of cancer patients increased by a great amount [19].

It should be mentioned that, at the beginning of this century, Japanese researchers were able to induce cancer in birds by staining them with coal tar [20]. This result led later to the discovery of a number of different carcinogens contained in this substance by Swiss, German, English, American, and Russian scientists (for references see [5]).

In the 20th century, cancer research has made large steps forward. Researchers have discovered several hundreds of chemical carcinogens and have found out in many cases the biochemical reactions they undergo in the cell before reaching their active, "ultimate" form which binds to DNA or protein constituents. It has been also found that different forms of radiation, e.g. ultraviolet (UV) and so-called ionizing radiation (high energy electromagnetic waves like X or γ rays, as well as particle radiation) can also cause cancer mostly by breaking the two strands of DNA in the same neighborhood. This causes a loss of a part of the genetic information. This effect of radiation can be understood only on the basis of physics and physical chemistry which underlie the science of radiology.

With the discovery of the human oncogenes (cancer causing genes which are dormant in our cells) the doors have been opened to understanding the physico-chemical and physical levels (on the levels of the electrons) how chemical carcinogens binding to DNA or radiation hitting them can activate these oncogenes or inactivate antioncogenes (cell duplication hindering genes). This results in a disturbance of the self-regulation of the cell (which can be described with the aid of appropriate mathematical methods). Thus it looks as if to prevent cancer, one needs the cooperation of nearly all the sciences ranging from medicine through biology and chemistry to physics and mathematics.

To make the summary of our knowledge about the onset of cancer in a biological cell understandable, we first of all have to describe briefly what a cell looks like. As everybody knows all living organisms are built up of cells. The cells of unicellular organisms (bacteria, certain algae etc.) are relatively simple (the so-called prokaryotic cells), while the different cells of higher multicellular organisms (the so-called eukaryotic cells) are rather complicated.

The prokaryotic cells are much smaller than the eukaryotic ones and they have only one mebrane which surrounds them:

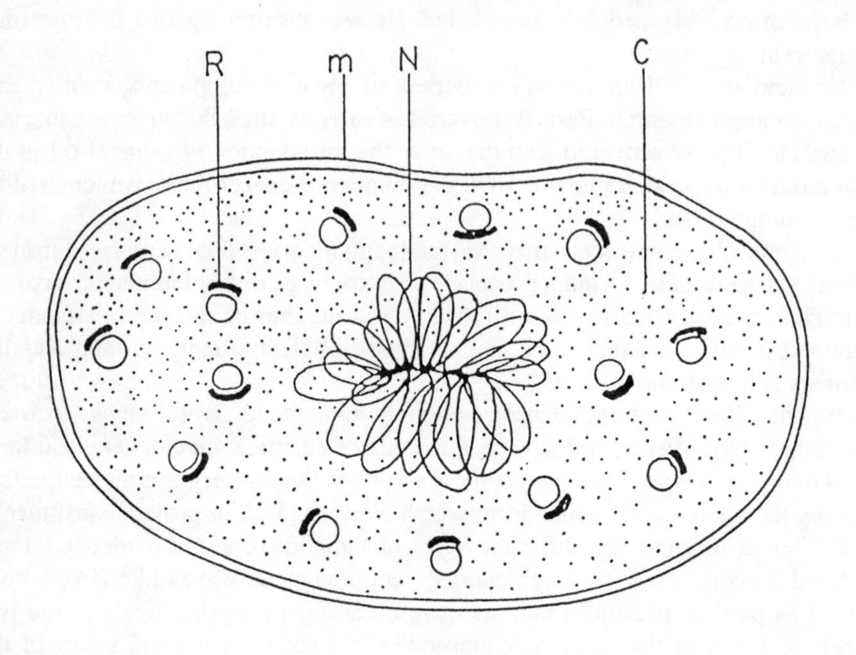

Fig. 1: Schematic representation of the cell of an unicellular organism (prokaryotic cell). *m* stands for the cell membrane, *N* for the nuclear zone (DNA), *N* for the ribosomes, and *C* for the cytoplasm

They contain only one chromosome which consists of a single molecule of double helical DNA, 0.12 cm in length [21] which is tightly coiled forming the so-called nuclear zone (prokaryotic cells have no nucleus). Besides this chromosome prokaryotic cells contain many ribosomes (in the best investigated bacterium, *Escherischia coli*, there are about 15000 ribosomes) where the protein synthesis takes place. Between the cell walls, the chromosome and the ribosomes, respectively, there is a very viscous aqueous solution containing about 20% dissolved proteins. This solution is called the cytoplasma. Prokaryotic cells have no oxygen metabolism, but only fermentation and they reproduce by asexual division.

The eukaryotic cells are much larger and more complicated than the prokaryotic ones (their volume is by a factor between 1000 and 10000 larger). They contain besides the cell membrane a membrane-enclosed nucleus [21]:

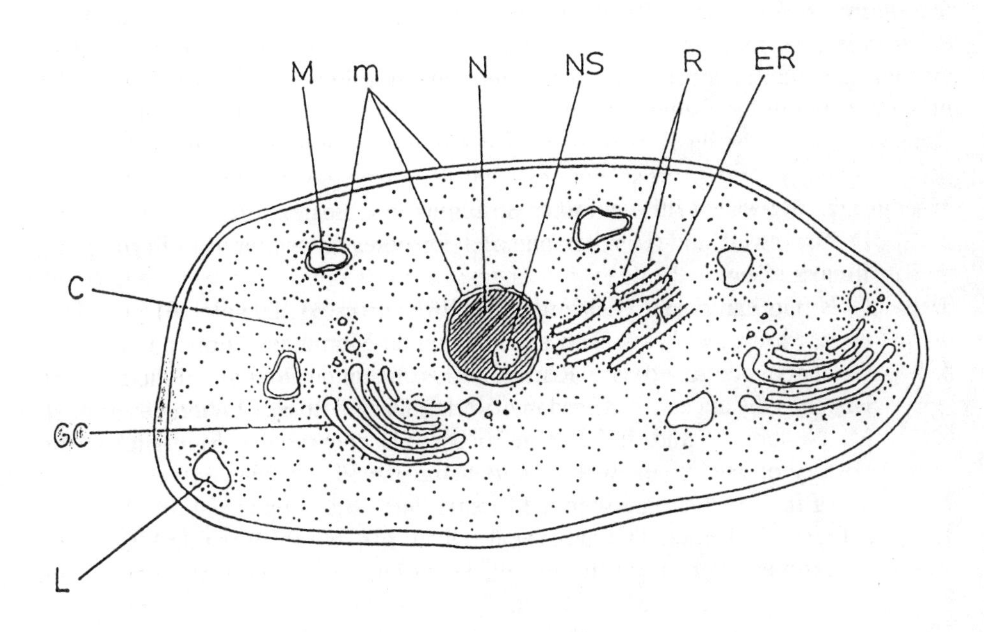

Fig. 2: Schematic representation of a cell of a multicellular organism (eukrayotic cell). *m* stands again for the membranes and *C* for the cytoplasm. *N* indicates the cell nucleus, *Ns* the nucleolus inside it, *M* stands for mitochondria, *L* for the lysosomes, *G* for the Golgi bodies, and finally *ER* for the endoplasmic reticuli (the ribosomes attached to it are shown as black points on the surface of the E.R. and they are indicated by R)

The DNA molecules in this nucleus are divided into several or many chromosomes which undergo mitosis at cell division. Inside the cell nucleus there is an even denser region which is called the nucleolus. Eukaryotic cells also contain other organelles surrounded by their own membranes. The oxygen metabolism of these cells takes place in the mitochondria, the Golgi bodies secret different cell products like proteins and help to construct certain membranes and the lysosomes digest different materials brought into the cell. The proteins synthesized in the ribosomes which are attached to the endoplasmic reticuli move in the cell along the channels formed by these single-membrane vesicles (see Fig. 2). The space between the cell membrane and all these organelles is filled again by the cytoplasm. Eukaryotic cells have an oxygen metabolism (though cancerous eukaryotic cells revert to fermentation), and in higher organisms they reproduce through genetic recombination (sexual reproduction, see Chapter 4).

References

1. R. L. Moodie, "The Antiquity of Disease", Chicago, 1923.
2. D. Brothwell and T. A. Sandison, "Diseases in Antiquity", Springfield, Illinois, 1968.
3. C. N. Goodman and G. M. Morant, Biometrika **31**, 295 (1940).
4. G. E. Smith and W. R. Dawson, "Egyptian Mummies", London, 1924.
5. G. Ebers, mentioned by J. Körbler, "Geschichte der Krebskrankheit" ("The History of Cancer", in German), Verlag Dr. H. Renner, Wien 1973, p. 6.
6. M. Neuburger and J. L. Pagel, "Die Geschichte der Medizin" ("The History of Medicine", in German), Jena, 1902.
7. The Bible, II. Chron., Sections 21 and 24.
8. J. Preuss, "Biblisch-Talmudische Medizin" ("Biblical-Talmudic Medicine", in German), Berlin, 1911.
9. I. Buhac, Deutsche Med. Woch. **88**, 287 (1963).
10. See for instance: F. A. Wise, "A Commentary on the Hindu System of Medicine", London, 1860.
11. "Herodoti Historiarum Libri IX", ed. H. R. Dietsch, Leipzig, 1894.
12. Hippokrates, "Collected Works", published (in French) by E. Littre, Paris, 1839-1861.
13. C. Tacitus, "Sämtliche Werke" ("Collected Works", in German), 1864; A. E. T. B. Wegall, "Nero, the Singing Emperor of Rome", London, 1930.
14. Galen, "Oeuvres de Galin" ("The Works of Galenus", in French), Paris, 1856.
15. Oreibasos, "Oeuvres d'Oribase" (by Boussemaker and Daremberg), Paris, 1851-1876.
16. Avicenna, "Kânûn", Rome, 1593.
17. Paracelsus, "Collected Works" (in German), ed. B. Aschner, Jena, 1926-1930.
18. See in Köbler's book (Ref. [5]), p. 67.
19. See in Köbler's book (Ref. [5]), p. 140.

20. K. Yamagiwa and K. Ischikawa, Proc. Med. Fac. Tokyo Univ. **15**, 255 (1916).
21. See for instance: A. L. Lehninger, "Biochemistry", Worth Publishers, Inc., 1975, pp. 30-33.

1 Do We Carry the Cause of Cancer in Our Genes?

In order to give an answer to this question we have first of all to describe very qualitatively the structure and function of the genetic material in a cell. More details will be given in Chapter 4 of this book and can also be found in appropriate text books, e.g. the book by Lehninger [1]. The genetic information of a cell is contained in giant molecules called deoxyribonucleic acid (DNA). The DNA molecules carry the information for the synthesis of the proteins necessary for cell functioning. The code for this information consists of four letters, the nucleotide bases guanine, adenine, thymine, and cytosine.

Each one of these nucleotide bases is bound to a sugar fragment, the deoxyribose. The sugar fragments are linked together by phosphate groups. In this way DNA forms a long chain which contains a certain sequence of nucleotide bases. The nucleotide bases form two complementary pairs: guanine can form three hydrogen bonds with cytosine and adenine two with thymine. In this way each strand of DNA is bound to a complementary one and these two chains form the famous double helix.

Each sequence of three nucleotide bases codes one of the twenty natural amino acids which build up the proteins. The complete sequence coding one protein is called a gene. For protein synthesis the double helix opens and a copy of the gene is made in form of a so called messenger ribonucleic acid (mRNA). This molecule contains as sugar ribose instead of deoxyribose, the nucleotide base uracil instead of thymine, and it forms no double helix.

In the next step transfer RNA (tRNA) molecules are bound to the mRNA. tRNA carries an amino acid at one end of the chain, and in its sequence the anticodon which is the group of the three complementary nucleotide bases coding the amino acid carried by the tRNA. With this anticodon the tRNA is bound at the complementary codon in the mRNA.

Finally the ribosomes, which catalyze protein synthesis, build covalent bonds between the amino acids at tRNA molecules bound to neighboring codons of the mRNA together to form the protein. The whole process is very complicated, but in order to be able to understand the following chapters this very qualitative overview is sufficient.

Besides the parts of the genes coding proteins, the so-called exons, the DNA also contains regulative sequences, which control the synthesized amount of a given protein, and sequences which mark starting and end point of a gene. Such regulative sequences are also found inside a gene and are called introns. These introns are missing in the sequence of the mRNA synthesized from a gene. Further sequences were found which, according to today's knowledge, have no function. The DNA double helices are bound by hydrogen bonds to special proteins, the nucleohistones, which give them a very complicated structure, forming the chromosomes. These nucleohistones also serve as blocking proteins and can prevent unnecessary or even dangerous information from being read (expressed). In later chapters we will see that these blocking proteins play a very important role in carcinogenesis.

In the next sub-chapter we will concentrate on the so called tumor viruses which are able to cause cancer. Although it is known that in human carcinogenesis the tumor viruses play a minor role (at least the retroviruses), their study had lead to the very important discovery of cancer causing genes in the genome of plants, animals and humans, the oncogenes.

1.1 Viral Cancer Causing Genes

Viruses consist in essence of their genetic material and the proteins which surround this material to form the viral particle. There are essentially two kinds of viruses which can cause cancer, the DNA viruses which contain DNA as genetic material [2] and the RNA viruses containing RNA [3]. Since the latter ones had been more important in the discovery of oncogenes we want to focus on them first.

The first suspicion that viruses might be one of the origins of cancer was mentioned by Peyton Rous in 1910 (for details and references see e.g. the papers by Bishop [4] and by Klein [5]) for the case of cancer in chicken caused by the Rous sarcoma virus. These kinds of viruses all act in the same way. They consist of a single stranded RNA molecule surrounded by proteins. After entering a cell and losing its proteins the virus uses one of its own enzymes, the reverse transcriptase to form a double stranded DNA-copy of its genetic material. This copy usually forms a ring molecule. Since the usual way of genetic information flow (DNA $\rightarrow$ RNA) is reversed in this case the RNA tumor viruses are also called retroviruses. The ring shaped viral DNA molecule is subsequently opened and integrated into the genetic material of the cell. Thus the cell now also produces mRNA copies of the viral genome.

The mRNA of viral origin is partly used to produce the proteins forming the virus particle and is partly incorporated into these new viruses. However, the RNA of the tumor viruses contains an additional gene, the viral oncogene. The expression of this gene by the cell leads to the synthesis of the so-called oncoproteins. As we will discuss later in more details, these might be proteins which are similar to proteins of the normal cell, but changed in some amino acids, or they can be normal cell proteins which are just overproduced. In the case of the Rous sarcoma virus the oncogene is named v-*src*, where v stands for its viral origin. The protein coded by this gene is an enzyme (a phosphorylase) which attaches a phosphate group to the tyrosine residues of proteins. The enzyme binds at the cell membrane and it is generally assumed that it changes the regulation of cell growth by this phosphorylation of membrane proteins. An interesting point is that many of these retroviruses, e.g. murine leukemia virus (MLV) are biologically not very active, but the new viruses produced by the infected cell are.

Further one should mention that not all retroviruses need to carry an oncogene to be able to cause cancer. All retroviruses have a nucleotide sequence, the LTR or "Long Terminal Repeat" at the ends of their RNA which, inserted into the sequence of cellular DNA causes a very high production rate of the viral proteins and can also lead to overproduction of cellular proteins. However, we will focus on this topic later.

Since the oncogenes of the more than 20 known tumor causing retroviruses all have similar counterparts in the cellular genomes (see next section), it is usually assumed that the retroviruses had taken up the oncogenes from the genomes of infected cells in the course of evolution [6]. Some examples for this fact are the *src* sequence of Moloney sarcoma virus which is similar to a normal sequence in mouse DNA [7] or the similarity of the protein sequences of histone H5 and the phosphoprotein p12 of MLV [8]. It is interesting to note that retroviral oncogenes are also able to cooperate in the initiation of cancer, i.e. the two oncogenes named ras and myc are far more active (higher percentage of cancerous cells produced in shorter time) when introuced together into cellular DNA than if one of them is introduced alone [9].

Finally it should be mentioned that the HIV virus which causes AIDS is also a retrovirus with some similarities to tumor viruses [10]. The connection between the viral oncogenes and cellular ones will be described in more detail below.

Let us turn now shortly to the other kind of tumor viruses, the DNA viruses. The most important ones of them are probably the hepatitis B virus which also causes liver cancer, the Epstein-Barr virus, causing e.g. lymphomas and some kinds of human papilloma viruses, which lead to genital cancers. It is believed nowadays that DNA viruses are one of the origins of about 20% of cancers which develop in the world [11]. It is assumed that the proteins produced from the viral DNA form complexes with, and thus inhibit the function of special proteins which normally control cell growth [11]. Some of the viruses also change the regulation of normal cellular genes by insertion of their genetic material close to the regulative sequences of these cellular genes. Further, the viral gene products can also cause mutations (changes of normal genes) in the cellular DNA which might lead to cancerogenesis. However, it is generally believed that none of the DNA tumor viruses cause cancer by themselves, but only together with other changes in infected cells. For a short review and further references see [12,13].

1.2 Oncogenes in Plants and Animals

After discussion of viral oncogenes let us turn now to cellular oncogenes, i.e. those which are already in the genome of cells i.e. without any virus infection. The observation that all viral oncogenes have very similar counterparts in the cellular genome naturally leads to the assumption that organisms like plants, animals and humans might already carry the origins of cancer in their normal genomes. From this, the question arises why these oncogenes have not been extinguished in the course of evolution and why genes contained in the normal genome can suddenly transform a normal cell into a cancerous one. First of all these genes found in normal DNA have to be called proto-oncogenes because they are not identical to the viral oncogenes but very similar. Thus they have to be activitated. This process will be discussed in Chapter 2 below. Secondly, the proto-oncogenes play usually an important role in the functioning (regulation) of normal cells and thus they could not be extinguished in the course of evolution.

The first oncogene discovered in animals is again the *src*-gene which was discussed above in connection with retroviruses. To distinguish the cellular *src* gene from the viral one the former is called c-*src* and the latter v-*src*. The existence of oncogenes was postulated on theoretical grounds as early as 1962 [14]. However, let us first turn shortly to oncogenes of non-viral origin in plants.

As just one example we want to concentrate on a kind of cancer which is observed in different kinds of plants (dicotyledonous), e.g. sunflower or tobacco [15], namely the crown galls. The interesting fact is, that these oncogenes are of bacterial origin, i.e. they are a unique example of the expression of prokaryotic DNA sequences by eukaryotes. The bacterium which is able to transfer carcinogenic plasmid DNA into the genome of plants is called *Agrobacterium tumefaciens* [16]. The plasmids which induce the cancer are ring shaped DNA molecules [17] and are different in different tumors, however, they contain as one part a special sequence of DNA which is the same in all cases and is called T-DNA [18]. Most probably this is the cancer causing sequence, i.e. the oncogene. The origin of cancer from the ring shaped so-called Ti-plasmids was proven by the observation that removal of this plasmid from oncogenic strains of *Agrobacterium tumefaciens* results in a loss of oncogenicity, while the introduction of this plasmid into the genome of non-oncogenic strains of the bacterium produces oncogenic forms [19,20]. Obviously we have here an example of an early discovered oncogene of non-viral but bacterial origin in plants. A review on this topic can be found in [21].

Let us turn now to non-viral oncogenes in animals, though, as we will see in a moment, the distinction between oncogenes in animals and humans is somehow artificial. As already mentioned, the first cellular oncogene discovered in animals was the c-*src* gene in chickens and other birds. One can identify these oncogenes by producing single stranded copies of the viral v-*src* gene which are radioactively marked and mixing them with single-stranded DNA of the normal cells. The result was the formation of hybrid (i.e. mixed) double-stranded DNA formed from viral and cellular DNA, i.e. in the normal cellular genome there has to be a gene similar (not necessarily identical) to v-*src*, because otherwise the two different kinds of DNA could not bind together (for details and further references on the *src* gene see [4,5]). Later the c-*src* gene was also found in the genome of mammals, including humans, and fishes. Therefore the distinction between oncogenes in animals and humans is somewhat artificial.

Later it was found that the c-*src* and the v-*src* genes are not identical. The c-*src* gene is in contrast to the viral one a so-called mosaic gene containing exons and introns. Due to the structural differences it is proven, that c-*src* is a typical cellular gene, and it is known that in normal cells c-*src* is also active. Since c-*src* is further contained in many kinds of animals on the ladder of evolution, its function for the normal cell must be very important. It was found that the products of v-*src* and normal c-*src* are very similar: both are enzymes which catalyze the phosphorylation of tyrosine and in both cases these enzymes are bound to the cell wall. Then the question arises why v-*src* is carcinogenic, while c-*src* in normal cells is not. The two hypotheses commonly accepted are either that the proto-oncogenes have to be modified in some way to become carcinogenic, or that an overproduction of the proteins coded by the oncogenes is responsible for their oncogenic activity. We will discuss in Chapter 2 the mechanisms of oncogene activation and we will see that both of the above mentioned possibilities are realized in nature, even sometimes for the same oncogene. As an example we want to mention here only that the injection of isolated cellular oncogenes (c-*src* and others) into normal cells lead to tumor formation, however, only when the cellular oncogenes were connected with a promoter gene from viral DNA. Thus for the oncogenic activity of c-*src* an overproduction of the coded protein is necessary, since otherwise no promoter gene would be needed in the above described experiments.

In the following we just want to mention some of the numerous cellular oncogenes which have been discovered in animals since the identification of c-*src*. A very interesting case is the oncogene v-*sis* (in simian sarcoma virus). It was found that the viral oncogene is similar to a gene occurring in the genome of the woolly monkey. In this case the protein product of the cellular counterpart is the so called PDGF (Platelet-Derived Growth Factor) polypeptide which stimulates the division of connective tissue cells and it seems that the *sis* oncogene turns the cells to malignancy by either overproduction of PDGF or by its production in an inappropriate time [22].

Another example of a cellular oncogene is the so-called c-*myc* gene. The viral counterpart v-*myc* is known to produce tumor viruses causing leukemia, sarcoma and carcinoma in chickens. The c-*myc* gene is found, as in almost all known cases of viral oncogenes, in several kinds of animals like chicken, rats and mice and also in humans. In rats it was found that a choline-free diet without adding carcinogens leads to amplification of the c-*myc* gene, i.e. to multiplication of the gene in the genome and consequently to overexpression, and to liver tumors [23]. The product of c-*myc* is again involved in growth control of cells, namely in the process of normal cell division and transformation [24]. Normal increase of c-*myc* production is found during liver regeneration [25] and also in some special kinds of (hepatocellular) carcinomas [26,27]. Interestingly it was found that many *myc* genes in tumor cells (Burkitt lymphomas) contain mutations at regions of the gene which are responsible for transcriptional attenuation which in the tumor cells is reduced [28] and there is evidence that this transcriptional attenuation is an important regulatory mechanism [29,30]. However, on the mechanisms of oncogene activation we will concentrate in more detail in Chapter 2.

In a study on the pre- and post-natal development of mice it was found that another oncogene, the Ha-*ras* oncogene, which is known from Harvey sarcoma virus, codes a protein which is found in virus infected cells as well as in normal cells [31] in a considerable but similar amount. It is expressed in mice embryos at any developmental states as well as in pre- and post-natal tissues at a similar level of activity. Thus the role of the c-*ras* product remains obscure, although activity in cell proliferation is possible because it was shown that introduction of the c-*ras* gene linked with a long terminal repeat unit of retroviral origin into mice cells leads to an uncontrolled cell proliferation [32]. Another variant of this oncogene, the Ki-*ras*, initially known from the Kirsten murine sarcoma virus, was found to cause in activated form lung and liver tumors in mice. These tumors occurred spontaneously as well as being chemically induced [33]. The *ras* oncogenes in activated form were observed also in several types of benign tumors in mice [34-36] (and also in human benign colon tumors [37]) and their activation seems to be one of initial steps of carcinogenesis in these cases.

In another type of mouse tumors, the plasmacytoid lymphosarcomas, the appearance of proteins coded by the so-called *myb* gene which are unusually larger than the products of this gene in normal cells results from a DNA-rearrangement in one chromosome at the site of the *myb* gene. Also this rearranged gene remained active in these tumor cells. The interesting fact is that the initial steps of carcinogenesis in these tumors require activation of another oncogene, named *abl*, but the *myb*- and also *myc*-products seem to be important for the retention of malignancy also in absence of the *abl*-products [38].

The *kis*-oncogene was found to be responsible for fibrosarcomas in mice. Also the c-*kis* gene is closely related to the retroviral v-*kis* gene of Kirsten murine sarcoma virus. In this case it was clearly found that the activated c-*kis* gene does not differ from the v-*kis* gene, thus it was established that the activation of the c-

kis gene is not due to any rearrangement of the DNA sequence of the gene. Therefore it seemed that c-*kis*, a member of the *ras* family, is most probably a case where the activation of the proto-oncogene results from an overexpression of the gene [39]. However, later in more detailed studies of the nucleotide base sequence of v-*kis* it was found that also point mutations may play a role in its activation.

A very interesting proto-oncogene, which will be discussed also in Chapter 2 is the *neu*-proto-oncogene, because it is known that the activation of this gene to become an oncogene requires the change of just one nucleotide base from adenine to thymine in its sequence, causing a change of valine to glutamic acid in the amino acid sequence of the protein encoded by the *neu* gene [40]. The neu proto-oncogene is similar but not identical to a gene which codes the epidermal growth factor receptor (EGFR) [41-43]. It is interesting that the human *neu* gene is more active than the rat *neu* gene, because the human variant becomes activated by overexpression [44,45] while the rat analogue does not.

To show the extent of conservation of cellular oncogenes in the course of evolution, one last remark seems to be in order. This is, that the vertebrate oncogenes have even been found in invertebrates, demonstrating both their high degree of conservation during evolution as well as their wide spread distribution among many divergent species. Namely, five different proto-oncogenes of vertebrates have been identified also in the fly *Drosophila melanogaster* [46]. Together with the wide-spread occurrence of these oncogenes in vertebrates this indicates again that proto-oncogenes fulfill functions in normal cells which are essential for their survival. Further these findings agree with the generally accepted notion that the viral oncogenes described in Section 1.1 are initially of cellular origin and were taken up by the retroviruses in modified form during infections in the course of evolution. Thus the possibility of a cell becoming malignant is obviously first of all carried by the genes of the normal cells and only subsequently is the cancerous information taken up by retroviruses.

In recent years, a very large number of cellular oncogenes have been observed and described. However, it does not seem to be necessary to list here all of them or to describe their properties. Instead we will give a short list of some retroviral oncogenes which have also been found in the genomes of normal animal cells. This list and further, more specialized information can be found in [47]:

Oncogene	Animal	Tumor type
v-*src*	Chicken	Sarcoma
v-*fps*	Chicken	Sarcoma

Oncogene	Animal	Tumor type
v-*yes*	Chicken	Sarcoma
v-*ros*	Chicken	Sarcoma
v-*myc*	Chicken	Carcinoma, Sarcoma, Leukemia
v-*erb*	Chicken	Sarcoma, Leukemia
v-*myb*	Chicken	Leukemia
v-*rel*	Turtle	Lymphoma
v-*mos*	Mouse	Sarcoma
v-*bas*	Mouse	Sarcoma
v-*abl*	Mouse	Leukemia
v-*ras*	Rat	Sarcoma, Leukemia
v-*fes*	Cat	Sarcoma
v-*fms*	Cat	Sarcoma
v-*sis*	Monkey	Sarcoma

The protein products of oncogenes can be partitioned into four main groups: (i) Growth factors which are bound to the cell membrane, (ii) Proteinkinases which are enzymes and catalyze the phosphorylation of tyrosine, serine or threonine side chains of proteins, (iii) GTP (guanosinetriphosphate) binding proteins which are involved in the transmission of extracellular signals through the cell membrane, (iv) Proteins in the nucleus of the cell which are most probably involved in regulation of gene expression and DNA replication. With the naturally incomplete list of examples above we want to turn to human oncogenes.

1.3 Human Cancer Causing Genes

As already discussed in the preceding Section, the cellular oncogenes identified in animals can all be found most probably also in the human genome. One of the first discoveries of a human proto-oncogene over a decade ago was the normal human cell gene homologous to the v-*ras* oncogene of Harvey murine sarcoma virus [48]. The c-*ras* human proto-oncogene is very similar to the viral one, however, in normal cells the concentration of the protein (p21) coded by this gene is much lower than in transformed ones. The human *ras* gene was connected to a long terminal repeat sequence of viral origin and inserted into the genome of mice where it caused cancer [48]. Even before this work it was known that DNA from malignant human cells is able to cause oncogenic transformation of normal cells [49-51]. These parts of the DNA sequences had been subsequently cloned and it was shown that also the cloned products were still oncogenic. Thus the genes themselves were the oncogenic agents and no external factor was necessary in this case [52,53]. Subsequently the cancer causing gene was identified as the human c-Ha-*ras*1 gene [54]. From these discoveries the existence of proto-oncogenes in the human genome was established. In the following decade more than hundred human oncogenes were identified. In fact nearly all cellular oncogenes found in animals so far should have their counterparts in the human genome, too. However, as discussed above, the genes in the normal cells are proto-oncogenes which have to be activated in some way to become carcinogenic. The possible mechanisms of this oncogene activation is the topic of the next chapter. Here we want to present and shortly discuss some further examples of human oncogenes. As in the case of oncogenes in animals, we want to concentrate here also on the findings published in the 1980s. For more specialized information, including complete lists of the human oncogenes discovered so far we refer the reader to the more recent literature.

The human proto-oncogene c-*bas*, being also a member of the *ras* family was investigated in thc same year [55] as the *ras*-gene. Also this gene was known before as a retroviral (murine sarcoma virus) oncogene v-*bas* which leads to bladder carcinoma. As mentioned above, by molecular cloning techniques a transforming gene could be isolated from human bladder carcinoma (T24 and EJ) cells [52,53, 56-58]. It was established that the oncogene is of human origin and that it is of a comparatively small size of less than 6000 nucleotide base pairs. It was found that the human analogue, c-*bas* in normal cells is indistinguishable from the oncogene isolated in the tumor cells by conventional techniques (enzyme analysis techniques). Also antisera which are known as being able to detect the protein products (oncoproteins) of the viral oncogenes could recognize the product of the human c-*bas* gene when it occurs in higher concentration than usual in normal cells [55]. As we know from the above considerations, the reason for that is, that for oncogene activation rather small alterations in the nucleotide base sequence of the proto-oncogene, or overproduction of the encoded protein can be sufficient. In the case of the c-Ha-*ras* proto-oncogene it was found for example that the difference to the oncogene from EJ bladder carcinoma cells consists of just one base, changing in the corresponding protein glycine to valine [59]. A similar situation is found in the human c-Ki-*ras* proto-oncogene, where in the corresponding activated oncogene probably again just one nucleotide base is substituted by another one [60]. At this point one should mention that the members of the *ras* family of human origin are also abbreviated as *HRAS*. Genes of the *ras* family encode cytoplasmic proteins. Note, that the above described experiments have mostly been performed in vitro or with special cell types which were already made immortal. Such so-called established cell lines already have a premalignant phenotype very similar to that induced by the *myc* oncogene. According to our present [61] knowledge it seems that no oncogene can cause cancer, when it acts alone on a completely normal cell. *ras*-like oncogenes can initiate cancer when acting alone only in immortal cells, or in cells which lack their normal environment which functions as a barrier against tumor growth. On the other hand, *ras*-like oncogenes together with *myc*-like ones can cause cancer also in completely normal cells (see [61] and references therein).

The c-*myc* oncogenes in animals as well as in humans encode nuclear proteins instead of cytoplasmatic ones [61]. An overactivation of the c-*myc* gene is observed in many different kinds of tumors, especially in colorectal carcinomas [62-64]. This overactivation is mostly connected with gene translocation, rearrangement or amplification (see Chapter 2 for details of these mechanisms). The human c-*myc* proto-oncogene is located within a DNA fragment of 12500 nucleotide base pairs [65]. In experiments on tumor cells from patients suffering from different kinds of cancer, it was found that in some cases rearrangements of the gene appeared, however, in one case the gene was not changed [66]. Further the activated c-*myc*-oncogene is able to immortalize normal cells, a fact which is interesting in connection with the finding discussed above that at least two oncogenes acting together should be necessary to initiate cancer. Since the cases where the *ras*-oncogene alone caused cancer were mostly observed in vitro with immortalized cells, one is tempted to assume that prior to the action of the *ras*-like genes, *myc*-like ones were already active to immortalize the cells [61].

Other human proto-oncogenes, c-*jun*, *jun*-B, *jun*-D, and c-*fos* encode proteins which are active in the nucleus and are parts of the mechanisms which convert stimuli from outside the cell into changes in the expression of genes [67]. Both oncogene types are also known from animals and retroviruses. Further, the encoded proteins form complexes with DNA in a very sequence-specific fashion, i.e. they bind only to specific nucleotide base sequences. Specifically, the *jun* protein binds to DNA while the *fos* protein does not build complexes with DNA directly but with the *jun* protein when it is bound to DNA [68]. The major protein in the transcription factor AP-1 is encoded by c-*jun*. It was further found that c-*jun* and c-*fos* oncogenes obviously act cooperatively.

An oncogene where no genetic alterations (mutations) were found after activation is the c-*met*. This proto-oncogene encodes a trans membrane tyrosine kinase [69,70]. This is known to be a receptor for a polypeptide which acts as a growth factor [71,72] and thus might have a strong influence on cell growth if its normal functioning is disturbed. The c-*met* oncogene was found to be amplified and overexpressed in human gastric carcinoma and in a colon carcinoma. In these cancerous tissues the receptor coded by c-*met* was found to be overexpressed between ten- and hundred fold [73].

It does not seem to be necessary to list here all the numerous human oncogenes which are known up to date. Their number already exceeds 100. About 30 of the human oncogenes can be found in the review by Aaronson [74]. A review on the proteins coded by oncogenes is contained in [75]. The discussion of the limited number of examples given above should be enough to get an idea of the basic principles. At least at this point we can answer the question given in the headline of this paragraph: "Do we carry the cause of cancer in our genes?" with yes. We have seen that the proto-oncogenes in our genome cannot cause cancer by themselves, but are all important for the normal functioning of our cells. Mostly their products are involved in regulation of cell growth and proliferation. Further we have seen that these proto-oncogenes have to be activated by external factors that they start their carcinogenic activity. This activation can consist in rearrangement (mutations) of the genes, in overproduction of the gene products or even in translocations of complete genes to another part of the genome where they are subsequently overexpressed again. Another, however, only artificial possibility for oncogene activation is the insertion of nucleotide base sequences called long terminal repeat (LTR) of viral origin at the starting point of the oncogene sequence. These mechanisms of oncogene activation will be discussed in some more detail in the following chapter. Further we have seen that in most cases one oncogene activated alone cannot cause cancer, but has to cooperate with one or more others.

Now we want to turn to another type of genes, which came into the discussion more recently, the so-called antioncogenes. The idea of this concept comes from the notion that instead of overactivation of a proto-oncogene also the reverse case might be possible, i.e. there might be regulatory agents which hinder the expression of oncogenes to a dangerous degree, and the beginning of carcinogenesis might be the inactivation of the genes which code these regulators. The idea emerged from the observation that the hybrid cells resulting from fusion of tumor cells with normal cells often loose their carcinogenicity [76-79]. This finding is in agreement with the concept that in the tumor some genes which produce repressors against overactivation of oncogenes might be blocked or lost, because by fusion with normal cells the information for the synthesis of these repressors can be regained in the hybrid cells. These antioncogenes are also sometimes called tumor suppressor genes because they regulate normal cell growth [61]. In [80] the isolation of such a gene was reported which is able to suppress partially the malignant phenotype of a cell made cancerous by activation of the *ras*-oncogene.

An interesting candidate for this is the *Rb* gene whose lack predisposes to retinoblastomas and osteosarcomas. The retinoblastoma tumors require mutations in two distinct genes before tumor development is possible [81]. It was experimentally established that one of these mutations creates inactive genes [82]. Therefore it seems that we not only carry the cancer causing information in our genome, but also genes which suppress carcinogenic activities of the otherwise necessary proto-oncogenes. Thus, besides the activation steps of oncogenes, for carcinogenesis probably the step of inactivation of antioncogenes also seems to be necessary. However, up to now neither the mechanisms for the activation of oncogenes nor those of the possible inactivation of antioncogenes are clear on the molecular level. In the next chapter we will turn to some of these mechanisms in a descriptive way, while their interpretation on the physical level is the topic of the main parts of this book.

References

1. A. L. Lehninger, "Biochemistry", Worth Publishers, Inc. (New York 1975).
2. See for instance "The Molecular Biology of Tumor Viruses, Part 2: DNA Tumor Viruses", J. Tooze (ed.), Cold Spring Harbour, NY: Cold Spring Harbour Laboratory, 1984.
3. See for instance "The Molecular Biology of Tumor Viruses, Part 1: RNA Tumor Viruses", R. Weiss, M. Teich, H. Varmus, and J. Coffin (eds.), Cold Spring Harbour, NY: Cold Spring Harbour Laboratory, 1984.
4. J. M. Bishop, New England J. Med. **303**,675 (1980); Cell **23**, 5 (1981); Ann. Rev. Biochem. **52**, 301 (1983); Nature **316**, 483 (1985).
5. G. Klein, in "Advances in Cancer Research", Vol. 37, W. Weinhouse (ed.), Academic Press (New York, 1982).
6. F. Wong-Staal, R. C. Gallo and D. Gillespie, Nature **256**, 670 (1975).
7. M. Oskarsson, W. L. McClements, D. G. Blair, J. V. Maizel, and G. F. VandeWoude, Science **207**, 1222 (1980).
8. L. E. Henderson, R. V. Gilden and S. Oroszlan, Science **203**, 1346 (1979).
9. S. J. Compere, P. Baldacci, A. H. Sharpe, T. Thompson, H. Land, and R. Jaenisch, Proc. Natl. Acad Sci. **86**, 2224 (1989).
10. R. Gallo, F. Wong-Staal, L. Montagnier, W. A. Haseltine, and M. Yoshida, Nature **333**, 504 (1988).
11. J. L. Marx, Science **243**, 1012 (1989).
12. D. Grunberger and S. Goff (eds.), "Mechanisms of Cellular Transformations by Carcinogenic Agents", Pergamon Press (Oxford, New York, 1987). Contains several chapters about RNA and DNA tumor viruses.
13. A. J. Levine, Cancer Res. **48**, 493 (1988).
14. H. Bush, in "Biochemistry of the Cancer Cell", Academic Press (New York, 1962), pp. 292.
15. M. De Cleene and J. DeLey, Bot. Rev. **42**, 389 (1976).
16. W. B. Gurley, J. Gallis, and J. D. Kemp, in "Genome Organization and Expression in Plants", C. J. Leaver (ed.), Plenum Press (New York, London, 1979), pp. 481.

17. I. Zaenen, N. Van Larebeke, H. Teuchy, M. Van Montagu, and J. Schell, J. Mol. Biol. **86**, 109 (1974).
18. M.-D. Chilton, R. K. Saiki, F.-M. Yang, K. Postle, A. L. Montoya, E. W. Nester, F. Quetier, and M. P. Gordon, in "Genome Organization and Expression in Plants", C. J. Leaver (ed.), Plenum Press (New York, London, 1979), pp. 471.
19. N. Van Larebeke, G. Engler, M. Holsters, S. Van den Elsacker, I. Zaenen, R. A. Schilperoort, and J. Shell, Nature **252**, 169 (1974).
20. B. Watson, T. C. Currier, M. P. Gordon, M.-D. Chilton, and E. W. Nester, J. Bacteriol. **123**, 255 (1975).
21. J. Schell and M. Van Montagu, in "Genome Organization and Expression in Plants", C. J. Leaver (ed.), Plenum Press (New York, London, 1979), pp. 453, and other papers in this volume.
22. J. L. Marx, Research News, Science **221**, 248 (1983).
23. N. Chandar, B. Lombardi and J. Locker, Proc. Natl. Acad. Sci. USA **86**, 2703 (1989).
24. M. D. Cole, Ann. Rev. Genet. **20**, 361 (1986).
25. N. L. Thompson, J. E. Mead, L. Braun, M. Goyette, P. R. Shank, and N. Fausto, Cancer Res. **46**, 3111 (1986).
26. G. J. Cote and J. F. Chiu, Biochem. Biophys. Res. Commun.**143**, 624 (1987).
27. B. E. Huber and S. S. Thorgeirsson, Cancer. Res. **47**, 3414 (1987).
28. S. Wright and J. M. Bishop, Proc. Natl. Acad. Sci. USA, **86**, 505 (1989).
29. D. L. Bentley and M. Groudine, Nature **321**, 702 (1986).
30. U. Siebenlist, P. Bressler and K. Kelly, Mol. Cell Biol. **8**, 867 (1988).
31. R. Müller, D. J. Slamon, J. M. Trembley, M. J. Cline, and I. M. Verma, Nature **299**, 640 (1982).
32. D. DeFeo, M. A. Gonda, H. A. Young, E. H. Chang, D. R. Lowy, E. M. Scolnick, and R. W. Ellis, Proc. Natl. Acad. Sci. USA **78**, 3328 (1981).
33. M. You, U. Candrian, R. Maronpont, G. D. Stoner, and M. W. Anderson, Proc. Natl. Acad. Sci. USA **86**, 3070 (1989).
34. S. Reynolds, S. Stowers, R. Maronpot, M. Anderson, and S. Aronson, Proc. Natl. Acad. Sci. USA **83**, 33 (1986).
35. A. Balmain, M. Ramsden, G. T. Bowden, and J. Smith, Nature **307**, 658 (1984).
36. S. Stowers, P. Glower, L. Boone, R. Maronpot,S. Reynolds, and M. Anderson, Cancer Res. **47**, 3212 (1987).
37. B. Vogelstein, E. Fearon, S. Hamilton, S. Kern, A. Preisinger, M. Leppert, Y. Nakamura, R. White, A. Smits, and J. Bos, N. Engl. J. Med. **319**, 525 (1988).
38. J. F. Mushinski, M. Potter, S. R. Bauer, E. P. Reddy, Science **220**, 795 (1983).

39. A. Eva and S. A. Aaronson, Science **220**, 955 (1983).
40. M.-C. Hung, D.-H. Yan and X. Zhao, Proc. Natl. Acad. Sci. USA **86**, 2545 (1989).
41. A. L. Schechter, D. F. Stern, L. Vaidyanathan, S. Decker, J. Drebin, M. I. Green, and R. A. Weinberg, Nature **312**, 513 (1984).
42. A. L. Schechter, M.-C. Hung, J. Vaidyanathan, R. A. Weinberg, T. Yang-Feng, U. Franke, A. Ullrich, L. Coussens, Science **229**, 976 (1985).
43. C. I. Bargmann, M.-C. Hung and R. A. Weinberg, Nature **319**, 226 (1986).
44. P. P. DiFiore, J. H. Pierce, M. H. Kraus, O. S. Segatto, R. King, S. A. Aaronson, Science **237**, 178 (1987).
45. R. M. Hudziak, J. Schlessinger and A. Ullrich, Proc. Natl. Acad. Sci. USA **84**, 7159 (1987).
46. B.-Z. Shilo and R. A. Weinberg, Proc. Natl. Acad. Sci. USA **78**, 6789 (1981).
47. J. M. Bishop, in "Krebs-Tumoren, Zellen, Gene" ("Cancer-Tumors, Cells, Genes), V. Schirrmacher (ed.), Heidelberg: Spektrum der Wissenschaft Verlagsgesellschhaft, 1989, pp. 52.
48. E. H. Chang, M. E. Furth, E. M. Scolnick, and D. R. Lowy, Nature **297**, 479 (1982).
49. C. Smith, B, Shilo, M. P. Goldfarb, A. Dannenberg, and R. A. Weinberg, Proc. Natl. Acad. Sci. USA **76**, 5714 (1979).
50. T. G. Krontiris and G. M. Cooper, Proc. Natl. Acad. Sci. USA **78**, 1181 (1981).
51. M. Perucho, M. Goldfarb, K. Shimizu, C. Lama, J. Fogh, and M. Wigler, Cell **27**, 467 (1981).
52. M. Goldfarb, K. Shimizu, M. Perucho, and M. Wigler, Nature **296**, 404 (1982).
53. S. Pulciani, E. Santos, A. V. Lauver, L. K. Long, K. C. Robbins, and M. Barbazid, Proc. Natl. Acad. Sci. USA **79**, 2845 (1982).
54. L. F. Parada, C. J. Tabin, C. Shih, and R. A. Weinberg, Nature **297**, 474 (1982).
55. E. Santos, S. R. Tronick, S. A. Aaronson, S. Pulciani, and M. Barbacid, Nature **298**, 343 (1982).
56. J. Bubenik, M. Baresova, V. Viklicky, J. Jakoubkova, H. Sainerova, and J. Donner, Int. J. Cancer **11**, 765 (1973).
57. C. J. Marshall, L. M. Franks and A. W. Carbonell, J. Natl. Cancer Inst. **58**, 1743 (1977).
58. C. Shih and R. A. Weinberg, Cell **29**, 161 (1982).
59. C. J. Tabin, S. M. Bradley, C. I. Bargmann, R. A. Weinberg, A. G. Papageorge, E. M. Scolnick, R. Dhar, D. R. Lowy, and E. H. Chang, Nature **300**, 143 (1982).

60. N. Tsuchida, T. Ryder and E. Ohtsubo, Science **217**, 937 (1982).
61. R. A. Weinberg, Cancer Res. **49**, 3713 (1989).
62. K. Alitalo, M. Schwab, C. C. Lin, H. E. Varmus, and J. M. Bishop, Proc. Natl. Acad. Sci. USA **80**, 1707 (1983).
63. P. G. Rothberg, J. M. Spandorfer, M. D. Erisman, R. N. Staroscik, H. F. Sears, R. O. Petersen, and S. M. Astrin, Br. J. Cancer **52**, 629 (1985).
64. J. Stewart,G. Evan, J. Watson, and K. Sikora, Br. J. Cancer **53**, 1 (1986).
65. R. Dalla Favera, M. Bregni, J. Erikson, D. Patterson, R. C. Gallo, and C. M. Croce, Proc. Natl. Acad. Sci. USA **79**, 7824 (1982).
66. A. Russo, M. LaFarina, D. Romancino, G. Grasso, V. Bazan, M. Alberti, I. Albanese, and P. Bazan, Anticancer Res. **12**, 1808 (1992).
67. A. G. Papavassiliou and D. Bohmann, Anticancer Res. **12**, 1805 (1992).
68. P. A. Sharp, Cancer Res. **49**, 2188 (1989).
69. M. Park, M. Dean, C. S. Cooper, M. Schmidt, S. J. O'Brien, D. G. Blair, and G. F. Vande Woude, Cell **45**, 895 (1986).
70. S. Giordano, C. Ponzetto, M. F. DiRenzo, C. S. Cooper, and P. M. Comoglio, Nature **339**, 155 (1989).
71. L. Naldini, E. Vigna, R. Narsimhan, G. Gaudino, R. Zarnegar, G. K. Michanopoulos, and P. M. Comoglio, Oncogene **6**, 501 (1991).
72. D. P. Bottaro, J. S. Rubin, D. L. Faletto, A. M. L. Chan, T. E. Kmiecik, G. F. Vande Woude, and S. A. Aaronson, Science **251**, 802 (1991).
73. M. F. DiRenzo, R. P. Narsimhan, M. Olivero, S. Pretti, S. Giordano, E. Medico, P. Gaglia, P. Zara, and P. M. Comoglio, Oncogene **6**, 1997 (1991).
74. S. A. Aaronson, Int. NFCR Symp. Carcinogenesis, Washington, DC 1984.
75. P. Kahn and T. Graf, "Oncogenes and Growth Control", Springer, Berlin, Heidelberg, New York, London (1986).
76. H. P. Klinger, Cytogenet. Cell Genet., **32**, 68 (1982).
77. R. Sager, Adv. Cancer Res. **44**, 43 (1985).
78. H. Harris, J. Cell Sci. Suppl. **4**, 431 (1986).
79. G. Klein, Nature **238**, 1539 (1987).
80. R. Shaefer, J. Iyer, E. Iten, and A. C. Nirkko, Proc. Natl. Acad. Sci. USA **85**, 1590 (1988).
81. A. G. Knudson, Prog. Natl. Acad. Sci. USA **68**, 820 (1971); Cancer Res. **45**, 1437 (1985).
82. J. J. Yunis and M. Ramsey, Am. J. Dis. Child **132**, 161 (1979).

2 Activation of Cancer-Causing Genes (Biochemical Discussion)

We have seen that the so-called cellular proto-oncogenes are all normal genes which usually encode proteins necessary for the normal functioning of the cell. Thus these genes in normal cells do not have the ability to initiate a carcinogenic transformation of the cells. However, under certain circumstances they obviously do. Therefore there have to be some mechanisms which are able to activate these genes, i.e. to introduce some changes which transform the non-carcinogenic proto-oncogenes into oncogenes which subsequently cause cancer. In this Chapter we want to describe the most important known mechanisms of oncogene activation from a biochemical point of view. That means that we are going to describe what are the changes in the proto-oncogenes which make them carcinogenic and, if known, what biochemical consequences they have. It is, however, beyond the scope of this Chapter to give the detailed chemical and physical mechanisms which cause these changes. This is the topic of later Chapters. In the last Chapter we mentioned some of the main mechanisms of oncogene activation. Now we want to describe them in some more detail. There are three main types of activation mechanisms. First of all genetic rearrangements in the nucleotide base sequences of proto-oncogenes, secondly overactivation of genes, which causes them to produce the proteins encoded by them in much higher quantities than usual, and thirdly the deactivation of genes which normally suppress such overproduction.

2.1 Activation of Oncogenes Through Mutations

As already discussed in the last Chapter it happens frequently that the activation of cellular oncogenes is accompanied by changes in the genetic code, i.e. the sequence of the nucleotide bases contained in the normal gene. Since the cancerogenic transformation of a normal cell is a major effect, one would expect that such genetic rearrangements should be rather drastic. However, what is found experimentally is the contrary - that very subtle changes, i.e. in many cases point mutations, are sufficient to transform a proto-oncogene into an oncogene. Surprisingly enough, in many cases, just the change of one nucleotide base in the sequence of the proto-oncogene, a point mutation is enough to make it cancerous. This corresponds to the change of just one amino acid in the sequence of the encoded protein into another one.

We want to elaborate somewhat on the first such case discovered, while the more recent findings of point mutations of other proto-oncogenes will be sketched more briefly. It seems to be important to describe in some more detail the mechanisms involved and experiments on just one example, since it is superfluous to list in detail all the cases of point mutations found up till now. This first discovery of a point mutation responsible for oncogene activation was made in the oncogene of human EJ bladder carcinoma a decade ago [1,2]. The responsible oncogene is the Ha-*ras* gene which is also present in the Harvey sarcoma virus and probably one of the most investigated oncogenes of all. Here we want to describe in some more detail the pathway which finally lead to the discovery of the point mutation as cause of the activation of this oncogene.

First of all it was found that the normal cellular genes did not induce cancer in normal cell cultures, while the genes isolated from tumor cells did with a high rate. The c-*ras* gene was isolated from human as well as from rat DNA and in form of v-*ras* from the Harvey sarcoma retrovirus [3-10]. However, despite this very drastic difference in function no differences in the structures of the normal gene and the activated oncogene could be found at first. Rough mapping techniques of the nucleotide sequences indicated no differences in the range of 6600 nucleotide bases containing the transforming activity of the tumor DNA [11]. With the help of molecular clones of the two genes subsequently a finer mapping of the sequence was possible but the only difference found was a mutation outside the coding regions of the genes which could be identified as a polymorphism of this gene which does not show any influence on its function, i.e. it is a so-called functionally silent polymorphism [1]. Examples of such polymorphisms had also been reported in other cases [12]. Thus this difference could not be the origin of the activation of this oncogene.

Consequently the researchers looked for more subtle differences between the two genes. The possibilities were a change in the regulation part of the gene or a change in the protein encoding part. The former possibility would have led to the overproduction of a normal cell protein, the latter one to the synthesis of a protein which is slightly changed. The products of *ras* genes are known to be proteins of a molecular weight of roughly 21000 and thus are called p21. It was found that the amount of RNA synthesized from the *ras* gene, as well as the amount of p21 proteins in normal and tumor bladder cells were comparable [1], thus ruling out the possibility of a mutation in the regulation sequence of the proto-oncogene. On the contrary new p21 proteins were found in the tumor cells, characterized by a higher mobility in electrophoresis experiments.

To obtain further information, sequence parts of the proto-oncogene were translocated into the oncogene and conversely parts of the oncogene into the proto-oncogene by biochemical techniques. The description of these techniques goes beyond the scope of this book. Afterwards the carcinogenic activity of the new genes obtained in this way was tested. With the help of this kind of experiments a region of just 350 nucleotide base pairs in the oncogene could be identified, which when transferred to the proto-oncogene lead to its carcinogenic activation [1]. Subsequently the nucleotide base sequences of this part of the proto-oncogene and the oncogene were determined. It was found that the two sequences just differed by one nucleotide base in codon (group of three nucleotide bases encoding one amino acid residue) 12 of the gene. At this position the normal proto-oncogene contains guanine (G), while in the oncogene thymine (T) was identified. In the protein synthetized from the genes this corresponds to a substitution of the amino acid glycine at position 12 in the normal protein to valine in the oncoprotein [1]. Since glycine is the smallest amino acid of all the 20 naturally occurring ones, while valine is rather bulky this should correspond to a rather large structural change in the oncoprotein, e.g. it is known that glycine is the strongest α-helix breaker of all amino acid residues [13]. Since activities of proteins are to a far extent connected with their spatial structure this single substitution might cause a very large change in the biological activity of a protein. Indeed it was shown by theoretical calculations, that the substitution of valine for glycine changed the spatial structure of the protein drastically [14].

A comparison of the p21 protein sequences in the human c-Ha-*ras*, the rat c-Ha-*ras* and the v-Ha-*ras* of viral origin, revealed that the p21 proteins in all three cases are identical with the sole exception of position 12. Therefore this region was strictly conserved in the course of evolution. Both the human and the rat c-Ha-*ras* product contains glycine at this position as already mentioned above. The human oncoprotein, however, contains valine, while in the viral oncoprotein arginine is found in this position. Thus it is clear that the substitution of glycine in exactly this position of the p21 protein by any other more bulky amino acid residue is important for oncogene activation. This is true both in case of oncogene activation when the retrovirus took up the gene from the cellular rat gene during evolution and in case of human carcinogenesis.

This is further proven by another member of the *ras*-family of oncogenes, the v-Ki-*ras* oncogene from Kirsten murine sarcoma virus. The Kirsten oncogene is identical to the v-Ha-*ras* oncogene [15], again with one exception: the encoded oncoprotein of v-Ki-*ras* oncogene contains the amino acid serine at the same position 12 [16] where the v-Ha-*ras* oncogene contains arginine, the c-Ha-*ras* oncoprotein valine, while the normal cell protein from the c-Ha-*ras* proto-oncogene contains glycine. Although the structure of the cellular proto-oncogene of v-Ki-*ras* is yet unknown, it is highly probable that also in c-Ki-*ras* the amino acid glycine should be encoded in that position.

To summarize, the activation of the c-Ha-*ras* oncogene seems to consist of just one nucleotide base alteration G → T. This causes a substitution of the amino acid glycine to valine in the encoded protein p21, which in turn leads to a drastic change in the three-dimensional spatial structure of this protein. Obviously this change occurs at the active site of the protein and thus disturbs the interaction of p21 with its cellular target considerably. One should point out that there are other well known cases where substitution of a single amino acid residue in a protein causes drastic changes in its function and even in cellular physiology. One example of this is the substitution of glutamine to valine in haemoglobin which changes its solubility in erythrocytes, causing the sickle-cell syndrome [1]. However, one should note here, that point mutation is not the only known way for activation of the c-Ha-*ras* proto-oncogene. This was shown experimentally by connecting c-Ha-*ras* proto-oncogenes without any changes in them to viral long terminal repeat (LTR) sequences which strongly enhance the production rate of the encoded protein. It was found that c-Ha-*ras* connected to LTR also leads to cancerogenic activity of the unchanged proto-oncogene [8,17]. Thus the c-Ha-*ras* proto-oncogene can be activated by two completely independent mechanisms which are both equally efficient in producing an active oncogene (see also the next section).

However, the reasoning described above holds only for the c-Ha-*ras* oncogene from EJ or also from T24 bladder tumor cells. Subsequently it was shown that the c-Ha-*ras* oncogene from other tumors is activated by another point mutation [18]. In that work, tumor cells of type Hs242 from a human lung carcinoma were used. The activated oncogene, also belonging to the c-Ha-*ras* proto-oncogene family is called c-*bas/has* to relate it to its viral origins v-*bas* from a mouse retrovirus and v-*has* from a rat retrovirus. The determination of the nucleotide base sequences of the different genes showed that in the first exon consisting of 37 codons the Hs242 oncogene is identical to the c-*bas/has* (c-Ha-*ras*) proto-oncogene from normal human cells, while as shown above the T24 and EJ oncogenes contain a thymine instead of a guanine in codon 12 (GTC instead if GGC, where C stands for cytosine). Therefore the activation of the Hs242 oncogene could not have been due to the same point mutation, although the oncogenes of the Hs242 lung tumor and the EJ and T24 bladder carcinoma derive from the same cellular proto-oncogene called c-*bas/has* or also c-Ha-*ras*. In the following, we will use the name c-Ha-ras to be consistent with the text given above.

With the help of the same techniques as in case of the EJ bladder oncogene it was possible [18] to identify a region in the Hs242 oncogene corresponding to the second exon and a part of the third in the gene to be responsible for its carcinogenic activity. This region had a length of 450 nucleotide base pairs and, together with the first exon where the point mutation responsible for EJ bladder carcinoma is located, is part of the genetic code for the p21 protein. A detailed sequence analysis of this region of the oncogene and the proto-oncogene revealed that in case of the Hs242 oncogene in codon 61, adenine (A) (proto-oncogene) is substituted by thymine (oncogene), i.e. the codon CAG coding for glutamine is changed to the codon CTG coding for leucine. This substitution again will be accompanied by a larger change in the spatial structure of the p21 protein which will inhibit its normal functioning in the cell [18]. Therefore we see that the p21 protein just needs to be changed in one amino acid, which can happen in different sites of the protein (site 12 for EJ bladder carcinoma, site 61 for Hs242 lung carcinoma), to aquire carcinogenic activity. Further, as already mentioned also overproduction of the normal p21 protein is sufficient to make it carcinogenic.

As third and last example we want to discuss the *neu* oncogene which was mentioned in the last Chapter. It was initially found in rat neuroblastomas generated by injection of ethylnitrosourea, a certain chemical carcinogen. The *neu* oncogene is also called murine c-*erbB-2*, its human homologue *ERBB2*, *HER2*, human c-*erbB-2* or *TKR1*. It was found [19] that the *neu* proto-oncogene is again activated by a single point mutation. In this case again adenine is changed to thymine, however, in another codon, namely GTG is transformed to GAG, which means that in the encoded protein, called p185, valine is substituted by glutamic acid. The proto-oncogene product is the epidermal growth factor receptor (EGFR) [20-22] protein, where the oncogenic mutation occurs in the trans-membrane section of the protein. However, in addition to the change in the encoded protein by the mutation, it is known that many human tumor cells contain an amplified *neu* gene [23-25], suggesting that activation of the *neu* oncogene might be a two step process, i.e. amplification followed by mutation. Amplification means that the gene is multiplied several times in the same strand (see below for more details). It was established, that amplification of the *neu* proto-oncogene alone does not lead to carcinogenesis, while amplification with subsequent mutation of a small number of the amplified proto-oncogenes results in tumor formation [26]. This might imply that in cases where up to now only overexpression was made responsible for the activation of oncogenes, also a two step mechanism could be functioning with a mutation as second step after amplification. These findings hold for the rat oncogenes, while it was found that for the human *neu* proto-oncogene even overactivation alone is sufficient for carcinogenesis [27,28]. Thus it seems that the human *neu* gene is more potent than the rat counterpart. We feel that these three examples for point mutations are sufficient for the scope of this section since the description of more such cases would not provide additional basic information. We only want to note here that, up to 1984, 16 such cases were known [29]. However, we want to emphasize that in all the cases described a relatively small amino acid in a normal protein is substituted by a more bulky (mostly containing large chemical side groups) amino acid in the corresponding oncoprotein and therefore the structure of the protein is changed considerably. Consequently the protein will be inactivated, or might change its function considerably.

At this point some additional remarks seem to be in order. The finding of point mutations does not explain how these mutations might occur on the chemical level. However, it seems to be obvious that for the attack of any kind of mutagenic agent on the proto-oncogenes, first of all the proteins bound usually to DNA have to be removed. This point will be extensively discussed in later chapters of this book. The two step mechanism for oncogene activation described above, however, seems to imply that prior to the mutation of the proto-oncogene, it has at least in some cases to be overexpressed. Overexpression in turn can occur only after removal of the blocking proteins from the DNA section to be expressed. Further, since we know that in the DNA double helix guanine is always bound to cytosine and adenine to thymine one has to ask in which way can such mutations occur. One mutation mechanism was proposed by Watson, Crick and Löwdin [30]. This mechanism involves the chemical concept of tautomerism. Tautomers are two molecules which are identical with the exception of the position of one hydrogen atom shifted to another binding site. In this way guanine has a tautomer G^* where the hydrogen atom of an NH group is shifted to the oxygen atom of the neighboring C=O group. This tautomer G^* can be obtained from normal G by excitation which requires a very low energy [31,32]. If such an excitation happens during the replication of DNA it would lead to a mutation, because G^* would bind to thymine instead of cytosine. The same can happen for adenine which also has a tautomer A^* which binds cytosine instead of thymine. When the new strand is then further copied one obtains substitution of adenine for guanine in the original code and vice versa. However, this comparatively simple mechanism cannot explain $G \rightarrow T$ or $A \rightarrow T$ substitutions as required in the above examples of oncogene activation. But there is the possibility that in the replication process due to an error in the complementary strand an adenine molecule is incorporated instead of cytosine. Then in the following copy one would obtain thymine and thus the required $G \rightarrow T$ substitution. However, this mechanism would involve the formation of a G-A base pair which would not fit into the double helix structure, but can happen if this structure is distorted, e.g. by binding of a carcinogen close to the mutation site [33]. These mechanisms will be explained in more detail in the forthcoming Chapters.

In principle, one could think that also rearrangements larger than point mutations might play a role in carcinogenesis. However, there is not much reported on this possibility in the literature. One is tempted to assume that the probability to have two ore more distinct point mutations on different locations in the proto-oncogene might be rather small. We have seen that for activation of the *neu* proto-oncogene a mutation from thymine to adenine is required, which converts valine to glutamic acid. It is known from in vitro experiments that *neu* activation can also happen by substitution of valine to glutamine or aspartic acid [34], what is not observed in vivo. This might well be due to the lower probability of this event, because these two substitutions require the change of two nucleotide bases in the GTG codon for valine. On the other hand, it was found that in the human tumor cells T-CAR-1 from the adrenal cortex besides overproduction of the EGFR protein from the *neu* oncogene in addition proteins similar to EGFR but with roughly half its size occur (see [26]). This could be explained by a mutation which deletes a large part of the *neu* oncogene leading to the synthesis of the smaller proteins. This assumption is corroborated by the fact that the viral v-*erbB* or v-*neu* oncogene encodes a truncated form of the EGFR protein.

2.2 Overactivation of Some Normal Genes

Now we want to turn to the mechanisms which might lead to overactivation of normal proto-oncogenes and in this way to carcinogenesis. Here we want to concentrate mainly on two such mechanisms, namely the binding of proto-oncogenes to a viral control element and to the translocation of oncogenes. The other possibility of mutations in the control elements of the proto-oncogenes themselves we feel is covered by the above section.

2.2.1 Overactivation by DNA Segments of Viral Origin

In order to avoid misunderstanding, we want to emphasize that the DNA segment we want to discuss here, the LTR sequence, does not occur in normal human or animal DNA. It plays a role in retroviral carcinogenesis and it is a very useful tool in experimental studies, because if bound to normal genes it leads to overproduction of the proteins encoded by these genes. The LTR occurs naturally only in DNA synthesized from retroviral RNA, i.e. in the DNA of virus infected cells. Therefore it plays only a minor role in human carcinogenesis, since cancers of viral origin are rare in humans. The DNA produced from viral RNA has the following general scheme:

U3 - R - U5 - *gag* - *pol* - *env* - U3 - R - U5

Here *gag* is a gene from which a polyprotein is synthesized. From this subsequently four proteins are formed (p15, p12, p30, and p10). These proteins are important in the inner structure of the virus and are called group specific antigenes, from which the abbreviation *gag* is taken. The protein synthesized from the gene *pol* (RNA dependent DNA polymerase) is the enzyme which catalyzes the formation of the DNA copy from the viral RNA. This protein is already contained in the virus when it infects a cell. Finally, the gene *env* (from envelope) codes the proteins of the virus membrane. Now we come to the LTR, which is the U3-R-U5 sequence. Here U3 is a unique sequence of about 500 to 800 nucleotide bases which marks the so-called 3' end of the DNA of viral origin. R is a so-called repeat part which contains about 70 nucleotide bases and occurs on both ends of the viral DNA. Finally U5 is again a so-called unique sequence of about

100 nucleotide bases and marks the 5' end of the chain. Note that the LTR sequence U3-R-U5 occurs only in the DNA synthesized from the viral RNA, not in the RNA strand itself which has the general sequence

R - U5 - *gag* - *pol* - *env* - U3 - R

Thus the enzyme coded by *pol* not only transcribes the viral RNA to a DNA strand, but also rearranges the end-groups of the RNA into the LTR groups at both ends of the synthesized DNA strand. In the case of most retroviruses the oncogene is not present in the original virus infecting the cell, but among the offsprings of such a virus (e.g. Moloney MLV, MLV for murine leukemia virus) a new type of virus appears (in this case Abelson MLV) in which the genes *pol* and *env* are completely and *gag* partially deleted. Instead the Abelson MLV contains a modified gene from the cellular genome, the oncogene v-*abl*. This new virus cannot propagate itself alone, but needs the original Moloney MLV as helper for its propagation.

However, at this point we are concerned with the LTR part of the viral DNA. The LTR is a regulating sequence which assures that after its insertion in the cellular genome the viral part of the new DNA is expressed dominantly. The cellular RNA-polymerase-II synthesizes the viral RNA from a promoter within the U3 part of the LTR segment in a very high concentration. In this way the propagation of the virus is ensured. However, if the viral DNA, as Abelson MLV, contains an oncogene, this oncogene is also overexpressed to a large extent due to the activity of the LTR segments at the ends of the viral genome. This oncogene might be one which is already modified from its cellular counterpart or it might be identical, becoming carcinogenic, i.e. activated, just by the overexpression caused by the LTR. However, in natural carcinogenesis, LTR leads only to overexpression of viral oncogenes, not of cellular ones. It might be possible that an LTR segment built into the cellular genome after virus infection could also lead to overexpression of cellular genes in its neighborhood. However, to the authors' knowledge there is no case reported in the literature, that in natural carcinogenesis LTR activates oncogenes which are not of viral origin.

However, since LTR is a defined segment of the viral RNA and thus can be extracted and bound to other genes, which can then be introduced into normal cells, it is an important tool for experimental research. If the question whether a given cellular proto-oncogene can be activated only by overexpression, without any mutations or other changes is raised, it can be answered experimentally with the help of the LTR sequence. Subsequently the proto-oncogene bound to the LTR sequence is introduced into the normal cells where it is overexpressed due to the activity of LTR. With this technique it was shown that the above discussed c-Ha-*ras* proto-oncogene becomes activated not only by point mutations but independently also by overexpression. In this experiment the unmodified human c-Ha-*ras* proto-oncogene was ligated to the LTR sequence from murine or feline retroviruses and then normal murine fibroblasts (a special kind of cells) were infected with this LTR-c-Ha-*ras* combination. Subsequently oncogenic transformation of these cells together with a very high concentration of the above described c-Ha-*ras* product (p21) was observed [17]. Note, that normal cells contain a low concentration of the p21 proteins encoded by the *ras* gene family. If just the human c-Ha-*ras* gene without an LTR is introduced into the normal cells it does not lead to oncogenic transformation. It was also shown experimentally that the transformed cells obtained from this experiment contained the c-Ha-*ras* gene together with the LTR control element.

This experiment together with those discussed above revealed that the *ras* proto-oncogene can be activated with different mechanisms independently: (i) point mutations which can occur at different codons and (ii) overactivation which in normal cells might occur by other activation mechanisms than ligation to viral LTR as in the experiment. Further it had been shown that the *ras* oncogene leads to oncogenic transformation in several tissues only when in addition also the *myc* oncogene is activated [35]. Just to mention a second example in this context we refer to the mouse c-*mos* proto-oncogene, which is not expressed at all in normal cells, while v-*mos* from Moloney murine sarcoma virus is a potent oncogene [36,37]. However, again if the mouse c-*mos* proto-oncogene is ligated to an LTR element it is expressed to a high rate in normal mouse cell lines and induces oncogenic transformation [37,38].

Thus it is obvious that ligation of cellular proto-oncogenes to LTR segments is an important experimental tool for the determination of the possible activation mechanisms of these genes, although it plays only a minor role in human carcinogenesis since it is of viral origin. The above given results together with other examples of activation of cellular oncogenes by ligation with viral LTR [8,39-41] suggest that an important mechanism of activation of cellular proto-oncogenes is just overexpression of these genes. In turn, this implies that in normal cells transcriptional control mechanisms must be active which prevent these genes from being expressed at a high rate. In the course of carcinogenesis in such cases, these mechanisms have to be inactivated in some way. This might happen by binding of proteins to control elements in the DNA close to the oncogenes, by a release of blocking proteins normally bound to DNA at the location of those oncogenes or in other, up to now unknown ways. There is also the possibility that mutations not in the protein coding segments but in the control segments of some oncogenes might be responsible for overexpression of the latter ones.

2.2.2 Translocation of Genes

A review on oncogene activation by translocation can be found in [42]. Here we want to describe this important mechanism for the activation of oncogenes again with the help of a specific example, namely the c-*myc* proto-oncogene. The mechanism of gene translocation appears mainly in cancer types which originate from the so-called immune-cells, namely the B cells which produce antibodies. In such cells the gene segments which are responsible for antibody production must be highly active. If in such a cell an oncogene would be located close to such an active segment its expression would be highly overactivated. It was found that in case of the Burkitt lymphoma a translocation occurs where the end parts of two chromosomes are exchanged [43-46]. In most cases an oncogene is changing its place to the neighborhood of the control element responsible for the high expression rate of the genes encoding antibodies. In some rare cases, however, also the control segment is moved into the vicinity of the oncogene.

An antibody consists of two identical pairs of protein molecules which are bound together to form a Y-like structure. Each of these pairs consists of a large protein, the so-called heavy chain and a smaller one, the light chain. In each of these chains there is a constant region, which is the same in all antibodies and there exist only two different versions of this constant region in the light chains and ten of them in the heavy chains. The variable regions exist in many different versions, depending which kind of antibody is encoded. However, one specific B-cell can synthesize only one type of antibody. It was found that the genes for the heavy chains are located on the human chromosome no. 14 and for the light chain on chromosome 22 or 2, depending which one of the two possible constant regions of light chains is encoded (see [42] and references therein).

In 1972 [47] it was found that chromosome 14 from Burkitt lymphoma cells was longer than usual and the aberration appeared at the end of the chromosome. Later [48] it turned out that a piece from the end of chromosome 8 breaks off in this case and is bound to the end of chromosome 14. In turn a smaller end part of chromosome 14 was deleted and bound to the end of number 8. In some cases of Burkitt lymphoma cells also exchanges of chromosome end-pieces between chromosome 8 and 2, and between 8 and 22 were observed. Note that in all cases chromosome 8 is involved and that all other chromosomes are those on which the genes encoding antibody proteins are located. Further the breaking point of chromosome 14, when exchanging parts with 8, is exactly in the region where the heavy chain is encoded [49,50].

The human proto-oncogene c-*myc* is naturally the most probable candidate as the oncogene for Burkitt lymphoma since its viral counterpart v-*myc* causes B-cell lymphoma in chicken. Therefore a search for the location of the human c-*myc* gene in the chromosomes was started. For this purpose a radioactively marked

DNA sequence very similar to v-*myc* was used and it was investigated with which parts of the human genome this sequence combines. It was found that in normal human cells c-*myc* is located in chromosome 8, while in Burkitt lymphoma cells c-*myc* is found on the small part of DNA which was translocated from chromosome 8 to 14 [50]. Interestingly, similar translocations of proto-oncogenes to an antibody encoding chromosome were observed subsequently in mice myeloma.

The c-*myc* gene can be translocated to chromosome 14 in different ways, namely sometimes the whole gene is translocated, in other cases only two of its three parts [42]. However, the lost part of c-*myc* does not code for the protein product but most probably it contains regulatory elements. Further it was found that the protein produced by c-*myc* in the Burkitt lymphoma cells is the same as that produced in normal cells. From this one can conclude that the oncogenic activitiy of the activated c-*myc* gene does not result from any mutation in the protein encoding nucleotide base sequence, but only from the translocation. Since after this translocation c-*myc* is situated in chromosome 14 close to the genes for the heavy chains of antibodies, it is in the cancer cells in a region of the genome which is highly active, and thus the protein product of the activated c-*myc* oncogene is synthesized in much larger amounts than in normal B-cells (from chromosome 8).

To prove this, the content of messenger RNA in the cells produced from c-*myc* was investigated. This was done with hybrid cells formed from human Burkitt lymphoma and normal mice myeloma cells. Since it was experimentally possible to distinguish between the c-*myc*-mRNA from humans and mice, it was found that c-*myc* on chromosome 14 was highly expressed, while the same gene on the second, unchanged chromosome 8 only to a very small amount (see [42]). In the other two types of translocations mentioned above, where DNA-pieces of chromosome 8 are exchanged with those from the antibody protein encoding chromosomes 2 and 22, respectively, c-*myc* remains on chromosome 8 but parts of the antibody encoding genes are placed close to c-*myc* on chromosome 8. Also in these cases overexpression, and thus activation of human c-*myc* was found [42]. Therefore c-*myc* itself does not need to be translocated for its activation. It only needs to have parts of genes in its vicinity which are related to antibody production. It makes no difference if this happens by migration of c-*myc* to the location of the genes encoding antibody proteins, or by translocation of genes related to this into the normal place of c-*myc* in the genome. The enzymes involved in the translocation mechanisms are nowadays known to a large extent, however, the detailed biological mechanism which leads to these events is too complicated to be explained in the context of this book.

In another experiment it was shown that the translocated c-*myc* proto-oncogene of Burkitt lymphoma is suppressed in hybrid cells formed with mouse fibroblasts, i.e. cells which do not produce antibodies, while it is highly expressed in hybrid cells formed with mouse myeloma cells, i.e. cells which produce antibodies. Thus it is shown that the translocated c-*myc* has an oncogenic activity only in antibody producing cells, in which the parts of the chromosomes necessary for this production are highly active. This is due to genetic elements contained in regions of DNA called enhancers which enlarge the transcription activity of genes in their neighborhood, but not only their next neighbors but in larger regions of the chromosome.

Since the above described results were obtained, the occurrence of gene translocations were found in many other tumor cells, especially in the T-cells, the second important part of the human immune system. Also in other tumor types occurring in B-cells translocations to chromosomes were found, however, from other chromosomes than 8. Thus in these cases, the c-*myc* gene is not involved in carcinogenesis, but other, yet unknown oncogenes on the chromosomes 11 and 14. Further also the previously mentioned c-*abl* proto-oncogene is involved in translocations [42] related to the chronic myelogenous leukemia (CML). The chromosome involved is the so-called Philadelphia chromosome Ph^1, to which the c-*abl* proto-oncogene from chromosome 9 is moved. However, in this case the activation of c-*abl* consists not only of overexpression of the gene, but also rearrangements of the gene take place. Moreover the Ph^1 chromosome is not an original chromosome from normal cells, but a fusion product of regions from chromosome 9, containing c-*abl*, and from chromosome 22. Obviously the activation of c-*abl* in this case is the molecular event which underlies the formation of the Ph^1 chromosome. Thus in this case the mechanism is completely different from that described above for the Burkitt lymphoma and c-*myc* activation.

2.3 Other Biochemical Possibilities of Oncogene Activation

In this final Section we want to discuss briefly some other mechanisms of oncogene activation. Briefly, because they, at least the first two, have been described already in the preceding points, especially in the chapter on oncogenes.

2.3.1 Amplification

Amplification of a gene means that with the help of enzymes, a given sequence in a DNA is multplied, i.e. after the process the gene is not only once present in the DNA chain but several times. This can have mainly two different effects. First of all a gene which occurs more than once is more often expressed and we know that overexpression leads to oncogenic activity of proto-oncogenes in several cases. The second effect is, that a gene which occurs more than once in the genome has a higher probability of being attacked by mutagens and thus its mutation rate is higher than that of a single gene.

An early model was proposed by Pall [53]. He proposed that the initial event of carcinogenesis by amplification should be a mutation of a proto-oncogene causing a single tandem duplication of the gene. The detailed mechanism of this process is still not well understood [53]. After DNA replication two sister chromatids are present, both carrying the duplicated gene. In further cell division and DNA replication sometimes an unequal sister chromatid crossing over can happen which creates a DNA chain (chromatid) with three copies of the proto-oncogene. Such events are known to occur infrequently from genetic studies [53]. In following repetitions of the cell division and DNA replication process, from time to time, more unequal crossover events happen generating more and more copies of the proto-oncogene until its expression rate is large enough to cause the cancerous transformation of the cell. This model could explain the long latency period following an attack of carcinogens.

An example where amplification of a proto-oncogene leads to a higher mutation rate has already been discussed and involves the *neu* proto-oncogene which causes, in activated form, e.g. neuroblastomas [26]. As mentioned above the activation of the *neu* proto-oncogene consists a single point mutation. However, in the case of *neu* a two step mechanism was found. First the proto-oncogene is amplified, probably following the mechanism proposed by Pall [53], leading first of all to an overexpression of the normal protein encoded by the gene. This is known from the observation that the tumor cells contain DNA with the amplified, unchanged proto-oncogene [23-25,54-56]. However, the fact that after amplification the proto-oncogene exists in several copies of the gene leads to a higher mutation rate. This facilitates the second step of oncogene activation, namely the point mutation leading to production of the oncoprotein. After the point mutation occurs, the activated oncogene is itself subject of the amplification process.

As a last example, we want to mention the c-*myc* proto-oncogene which is amplified in liver tumors. However, the amplified proto-oncogene, and consequently an overproduction of the encoded protein was found also in non-carcinogenic cells. Thus here again the amplification of the c-*myc* gene is the first step of oncogene activation, which must be followed by a second step, probably mutations. However, the exact type of this second step in the kind of liver tumors studied is not yet clear [57]. The interesting point in this study is that the cancerous transformation was induced in rats by feeding them a choline free diet, containing no additional mutagenic carcinogens. Further the size of the amplification of c-*myc* is controllable by the experimental conditions, concerning the precise nature of the diet. Finally the choline free diet was the first experimental system where it was possible to introduce tumors with amplification in a reproducible way.

There are many other examples known where amplification of a proto-oncogene is one of the steps of carcinogenesis. However, for the understanding of the basic mechanisms it is not necessary to give a longer list of examples at this point.

2.3.2 Antioncogene Deactivation

The concept of antioncogenes was discussed in some detail in the point on human oncogenes. Of necessity, we also had to describe the mechanisms with which antioncogenes act at that point and thus we want to keep the discussion here short. A detailed description of the cooperation of antioncogenes and oncogenes can be found in the paper by Weinberg [58]. Let us take for instance a proto-oncogene like *ras* for which it is known that its overexpression without any mutations can lead to carcinogenesis. It is known that *ras* usually has to cooperate with an activated nuclear oncogene, the c-*myc* oncogene. Thus one is tempted to assume that instead of activating a proto-oncogene, it should be a far more easy process to deactivate another gene which regulates the expression rate of *ras* or *myc* in the normal cell (see [58] for details and further references). Since both *ras* and *myc* are also active in normal cells, there must be a suppressor gene which ensures that these proto-oncogenes are not overexpressed in the normal functioning of the cell. From many experiments it is known that it is in general far more easy to destroy gene functioning (the antioncogene) than to create highly activated versions of a normal gene (the oncogene). Experimental evidence for this idea is the fact that the fusion of normal cells with tumor cells often leads to hybrids which are no longer cancerous [59-63]. This proves that carcinogenesis must often involve the loss of some genetic information which can be reacquired by fusion with normal partner cells. The first such tumor suppressor gene or antioncogene, the so-called *Rb* gene (from retinoblastoma) is described already in the point on human oncogenes and thus we don't want to elaborate again on it here. However, what was not discussed there was the deactivation mechanism for this antioncogene. The *Rb* antioncogene encodes a nuclear phosphoprotein (p105-*Rb*) which is known to form complexes with a tumor virus (adeno virus E1A) and also with some oncoproteins (SV40 large T). It was discovered that in J82 bladder carcinoma cells a variant of this *Rb* protein is present which does not form complexes with E1A and which is also less stable than normal p105-*Rb*. It was established that for the formation of this changed protein in the J82 genome, one exon is discarded, causing the loss of 35 amino acids in the encoded protein [64]. The genetic reason for this is a point mutation in a splice acceptor sequence, changing a 5'-TTTATTTACTAG-3' sequence into 5'-TTTATTTACTGG-3' or, with other words, an A $\rightarrow$ G mutation in the last codon of this sequence. Since this mutation does not change the protein encoding part of the gene, as we have seen in cases discussed above, but the regulatory part, it leads to a rather drastic consequence: the exon 21 is discarded during expression of this gene, and exon 20 is directly fused to exon 22, leading to the above mentioned loss of 35 amino acids in the protein. In this way the protein is no longer able to function as a tumor suppressor, i.e. the antioncogene is deactivated by a simple point mutation, just as we have seen in the case of activation of oncogenes (*ras* and *neu*) by point mutations. Further information on the current state of research on antioncogenes and references can be found in [65]. As will be discussed below, e.g. double strand breaking of DNA caused by radiations can lead to the loss of antioncogenes, and thus in turn to their deactivation.

2.3.3 DNA Transposition

Another mechanism of oncogene activation which has not been mentioned so far was found for the c-*mos* proto-oncogene. Its viral counterpart (from Moloney murine sarcoma virus) v-*mos* contains 374 codons, encoding a protein with a molecular weight of about 37000. The cellular normal c-*mos* gene is 21 codons longer than the viral counterpart. In normal cells the protein encoded by c-*mos* was not found. Thus it seems that c-*mos* is inactive in normal cells. It was found in a mouse myeloma that in the activated form called rc-*mos* a DNA rearrangement occurred at 5'-end of the cellular gene. In this rearrangement c-*mos* information is deleted and an insertion sequence (IS) of 159 base pairs is added. This case was the first example of a non-virally induced tumor, where a cellular oncogene is activated by a DNA transposition [66]. There was evidence that the insertion of the IS element happens prior to the deletion of c-*mos* information. Further it is assumed that the new IS element activates the oncogenes through a strong promoter sequence contained in it. The new sequence shows no similarity to viral LTR sequences, but has in common several details with bacterial IS elements, i.e. (i) the sequence ends in a terminal repeat and (ii) the terminal nucleotide bases CA and TG are found in all eukaryotic cellular mobile elements. Such so-called "jumping genes" had been discovered [67] even before the connection of some special ones of them to carcinogenesis was found. This experimental finding was honored with the Nobel prize later.

References

1. C. J. Tabin, S. M. Bradley, C. I. Bargmann, R. A. Weinberg, A.P. Papageorge, E. M. Scolnick, R. Dhar, D. R. Lowy, and E. H. Chang, Nature **300**, 143 (1982).

2. E. P. Reddy, R. K. Reynolds, E. Santos, and M. Barbacid, Nature **300**, 149 (1982).
3. C. Der, T. G. Krontiris,and G. M. Cooper, Proc. Natl. Acad. Sci. USA **79**, 3637 (1982).
4. L. Parada, C. J. Tabin, C. Shih, and R. A. Weinberg, Nature **297**, 474 (1982).
5. E. Santos, S. R. Tronick, S. A. Aaronson, S. Pulciani, and M. Barbacid, Nature **298**, 343 (1982).
6. E. M. Scolnick and W. P. Parks, J. Virol. **13**, 1211 (1974).
7. T. Y. Shih, D. R. Williams, M. O. Weeks, J. M. Maryak, W. C. Vass, and E. M. Scolnick, J. Virol. **27**, 45 (1978).
8. D. DeFeo, M. A. Gonda, H. A. Young, E. H. Chang, D. R. Lowy, E. M. Scolnick and R. W. Ellis, Proc. Natl. Acad. Sci. USA **78**, 3328 (1981).
9. E. H. Chang, M. A. Gonda, R. W. Ellis, E. M. Scolnick, and D. R. Lowy, Proc. Natl. Acad. Sci. USA **79**, 4848 (1982).
10. R. W. Ellis, D. DeFeo, J. M. Maryak, H. A. Young, T. Y. Shih, E. H. Chang, D. R. Lowy, and E. M. Scolnick, J. Virol. **36**, 408 (1980).
11. C. Shih and R. A. Weinberg, Cell **29**, 161 (1982).
12. M. Goldfarb, K. Shimizu, M. Perucho and M. Wigler, Nature **296**, 404 (1982).
13. C. R. Cantor and P. R. Schimmel, "Biophysical Chemistry", Vol. 1 (Freeman, San Francisco, 1980), p. 303.
14. M. R. Pincus, J. van Ransoude, J. B. Harford, E. H. Chang, R. P. Carty, and R. D. Klausner, Proc. Natl. Acad. Sci. USA **80**, 5253 (1983).
15. T. Y. Shih, M. O. Weeks, H. A. Young, and E. M. Scolnick, Virology **96**, 64 (1979).
16. N. Tsuchida, T. Ryder and E. Othsubo, Science **217**, 937 (1982).
17. E. J. Chang, M. E. Furth, M. E. Scolnick, and D. R. Lowy, Nature **297**, 479 (1982).
18. Y. Yuasa, S. K. Srivasta, C. Y. Dunn, J. S. Rhim, E. P. Reddy, and S. A. Aaronson, Nature **303**, 775 (1983).
19. C. I. Bargmann, M.-C. Hung and R. A. Weinberg, Cell **45**, 649 (1986).
20. A. L. Schechter, D. F. Stern, L. Vaidyanathan, S. Decker, J. Drebin, M. I. Green, and R. A. Weinberg, Nature **312**, 513 (1984).
21. A. L. Schechter, M.-C. Hung, J. Vaidyanathan, R. A. Weinberg, T. Yang-Feng, U. Franke, A. Ullrich, and L. Coussens, Science **229**, 976 (1985).
22. C. I. Bargmann, M.-C. Hung and R. A. Weinberg, Nature **319**, 226 (1986).
23. D. J. Slamon, G. M. Clark, S. G. Wong, W. J. Levin, A. Ullrich, and W. L. McGuire, Science **235**, 177 (1987).
24. M. H. Kraus, N. C. Popescu, S. C. Amsbaugh, and C. R. King, EMBO J. **6**, 605 (1987).

25. M. Vijver, R. Bersselaar, P. Devilee, C. Cornelisse, J. Peterse, and R. Nusse, Mol. Cell Biol. 7, 2019 (1987).
26. M.-C. Hung, D.-H. Yan and X. Zhao, Prroc. Natl. Acad. Sci. USA **86**, 2545 (1989).
27. P. P. DiFiore, J. H. Pierce, M. H. Kraus, O. S. Segatto, R. King, and S. A. Aaronson, Science **237**, 178 (1987).
28. R. M. Hudziak, J. Schlessinger and J. A. Ullrich, Proc. Natl. Acad. Sci. USA **84**, 7159 (1987).
29. S. A. Aaronson, Int. NFCR Symp. Carcinogenesis, Washington DC (1984).
30. J. D. Watson and F. H. C. Crick, Nature **171**, 737, 964 (1953); J. D. Watson, "The Double Helix" (Atteneum, New York, 1968); P.-O. Löwdin, Rev. Mod. Phys. **35**, 724 (1963).
31. J. Ladik, J. Theor. Biol. **6**, 201 (1964).
32. R. Rein and J. Ladik, J. Chem. Phys. **40**, 2466 (1964).
33. J. Ladik, Intern. J. Quantum Chem. **26**, 955 (1984).
34. C. I. Bargmann and R. A. Weinberg, EMBO J. 7, 2043 (1988).
35. S. J. Compere, P. Baldacci, A. H. Sharpe, T. Thompson, H. Land, and R. Jaenisch, Proc. Natl. Acad. Sci. USA **86**, 2224 (1989).
36. M. Oskarsson, W. L. McClements, D. G. Blair, and J. V. Maizel, Science **297**, 1222 (1980).
37. D. G. Blair, M. Oskarsson, T. G. Wood, W. L. McClements, P. J. Fischinger, and G. G. Vande Woude, Science **212**, 941 (1981).
38. S. Gattoni, P. Korschmeier, I. B. Weinstein, J. Escopbedo, and D. Dina, Molec. Cell Biol. **2**, 42 (1982).
39. W. S. Hayward, G. B. Neel and S. M. Astrin, Nature **290**, 475 (1981).
40. G. S. Payne, S. A. Courtneidge, L. B. Crittenden, A. M. Fadly, J. M. Bishop, and H. E. Varmus, Cell **23**, 311 (1981).
41. G. B. Neel, W. S. Hayward, H. L. Robinson, J. Fang, and S. M. Astrin, Cell **23**, 323 (1981).
42. F. G. Haluska, Y. Tsujimoto and C. M. Croce, Ann. Rev. Gen. **21**, 321 (1987).
43. R. Taub, I. Kirsch, C. Morton, G. Lenoir, D. Swan, S. Tronick, S. A. Aaronson, and P. Leder, Proc. Natl. Acad. Sci. USA **79**, 7837 (1982).
44. G. Klein, Cell **32**, 311 (1983).
45. G. Klein and E. Klein, Immun. Today **6**, 208 (1985).
46. I. D. Rowley, Cancer Res. **44**, 3159 (1984).
47. G. Manolov and Y. Manolova, Nature **237**, 33 (1972).
48. L. Zech, U. Haglund, K. Nilsson, and G. Klein, Int. J. Cancer **17**, 47 (1976).
49. J. Erikson, J. Finan, Y. Tsujimoto, P. C. Nowell, and C. M. Croce, Proc. Natl. Acad. Sci. USA **81**, 4144 (1984).

50. J. Erikson, D. L. Williams, J. Finan, P. C. Nowell, and C. M. Croce, Science **229**, 784 (1985).
51. R. Dalla-Favera, S. Martinotti, R. C. Gallo, J. Erikson, and C. M. Croce, Science **219**, 963 (1983).
52. R. Dalla-Favera, M. Bregni, J. Erikson, D. Patterson, R. C. Gallo, and C. M. Croce, Proc. Natl. Acad. Sci. USA **79**, 7824 (1982).
53. M. L. Pall, Proc. Natl. Acad. Sci. USA **78**, 2465 (1981).
54. T. A. Libermann, H. R. Nusbaum, N. Razon, R. Kris, I. Lax, H. Soreq, N. Whittle, M. D. Waterfield, A. Ullrich, and J. Schlessinger, Nature **313**, 144 (1985).
55. K. Semba, N. Kamata, K. Toyoshima, and T. Yamamoto, Proc. Natl. Acad. Sci. USA **82**, 6497 (1985).
56. Y. H. Yu, N. Richert, S. Ito, G. T. Merlino, and I. Pastan, Proc. Natl. Acad. Sci. USA **81**, 7302 (1984).
57. N. Chandar, B. Lombardi and J. Locker, Proc. Natl. Acad. Sci. USA **86**, 2703 (1989).
58. R. A. Weinberg, Cancer Res. **49**, 3713 (1989).
59. H. P. Klinger, Cytogenet. Cell Genet. **32**, 68 (1982).
60. R. Sager, Adv. Cancer.Res. **44**, 43 (1985).
61. H. Harris, J. Cell. Sci. Suppl. **4**, 431 (1986).
62. G. Klein, Nature **238**, 1539 (1987).
63. A. G. Geiser, C. J. Der, C. J. Marshall, and E. J. Stanbridge, Proc. Natl. Acad. Sci. USA **83**, 5209 (1986).
64. J. M. Horowitz, D. W. Yandell, S.-H. Park, S. Canning, P. Whyte, K. Buchkovitch, E. Harlow, R. A. Weinberg, and T. P. Dryja, Science **243**, 937 (1989).
65. "Abstracts of the Fourth International Conference of Anticancer Research", Anticancer Res. Vol. **12** (1992).
66. G. Rechavi, D. Givol and E. Canaani, Nature **300**, 607 (1982).
67. See e.g.: B. McClintock, in "Brookhaven Symposia in Biology", Vol. 18, 162 (1965) for a review and references therein.

3 The Role of External Factors in the Start of the Cancerous Change in a Cell

There are several different kinds of external, cancer causing factors known, of which we want to discuss the most important classes in this chapter. First of all there are several hundred chemicals which are known to cause cancer, the so-called chemical carcinogens, which are dealt with in the first section. Secondly, ultraviolet (UV), γ-, X-, and particle radiation can cause cancer. Further the lack or in other cases the abundance of some trace elements in a cell can cause the malignant change. Finally we want to deal with chemicals, the cancer promoters, which by themselves are not carcinogens, but when they act together with carcinogens even in very small doses, they enhance the carcinogenic activity of the latter. Let us first turn now to the chemical carcinogens.

3.1 Different Cancer Initiating Chemical Compounds (Carcinogens) and Their Most Important Reactions in the Cell

Carcinogens were discovered as early as the 19th century in Britain. In 1822 the carcinogenicity of arsenic was discovered [1], and it was found that working with mineral oil in industry increased the number of cancer patients [2]. Later, in Japan, the carcinogenicity of coal tar was discovered [3]. Since then several hundreds of carcinogens have been identified and therefore we cannot discuss each of them in detail or even mention all of them. Thus we will cover a couple of representative examples in the following. Here we have to note first of all that carcinogens can be divided into two main groups. There are the directly acting carcinogens, which react with DNA as they are, and the indirectly acting ones. In the latter cases not the carcinogens themselves attack DNA but only their metabolized forms. Thus we have to discuss not only the carcinogens, but also their reactions which lead, after metabolism, to their so-called ultimates, the chemicals which really attack DNA. This process is also called enzymatic activation. After that we have to ask how these ultimates or the directly acting carcinogens react with DNA or in other words in which way are they able to activate oncogenes.

The directly acting carcinogens can again be divided into two distinct groups, the alkylating ones which substitute an alkyl residue for a proton and the non-alkylating group which undergoes other reactions with DNA. These non-alkylating agents are usually simple chemicals and as representative examples we want to mention nitrous acid: O=N-OH, hydroxylamine: H_2N-OH, methoxyamine: H_2N-O-CH_3, hydrazine: H_2N-NH_2, and formaldehyde: H_2C=O. These chemicals can attack and substitute amino groups which are outside the ring systems of the nucleotide bases, so-called exocyclic amino groups. This process is called deamination. On the other hand they can also shift the tautomeric equilibrium between the different forms of the nucleotide bases, as already described above. In both cases, they can change base pairing in DNA and thus lead to mutations and eventually also to activation of oncogenes. Nitrous acid for example can change cytidine (CS, S for the sugar residue) to uridine (US), adenosine (AS) to a deaminated form called HXS (Hydroxantin) and guanosine (GS) to a form called XS (Xantin) [4]:

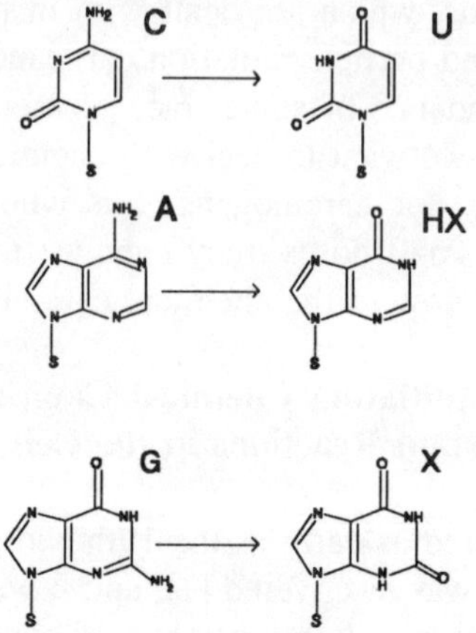

C → U

A → HX

G → X

Obviously this kind of reaction should have a drastic influence on base pairing. Another reaction type of nitrous acid is the formation of cross links between two GS or a GS and an AS group:

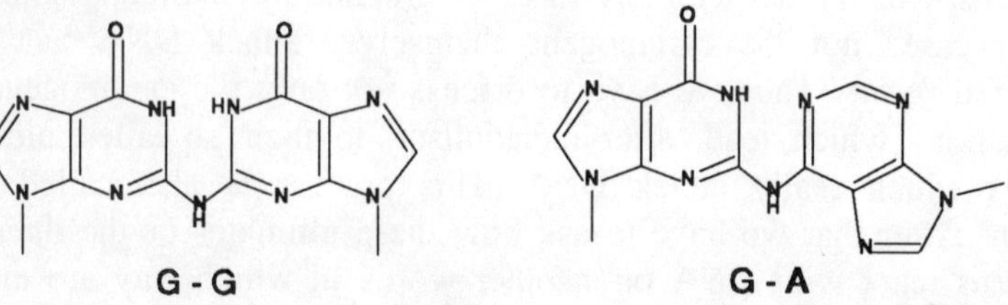

G - G   G - A

The cross links are thought to inactivate these bases in the sequence biochemically. Note, that nitrous acid as well as the other examples react predominantly with single stranded polynucleotides. The other groups of chemicals mentioned above also react with exocyclic amino groups of the nucleotide bases predominantly in single stranded chains. The two possible reaction pathways of hydroxylamine and methoxyamine with cytidine are first an addition of the chemical to the C=C double bond in cytosine, followed by deamination and a subsequent elimination of the first reactant [5]:

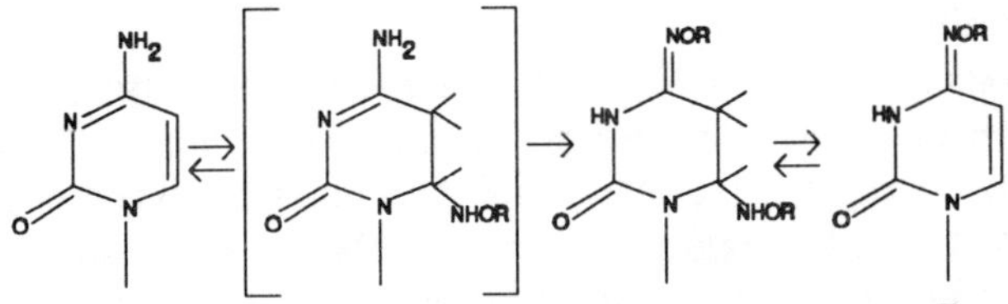

The second pathway is a direct deamination.

Hydrazine, which is used industrially to a great extent, reacts again with the exocyclic amino group (-NH_2) of cytidine, replacing it with -NH-NH_2 [6]. In a similar way formaldehyde is also attacked nucleophilically by these -NH_2 groups of cytidine, adenosine and guanosine, forming -NH-CH_2-OH groups [7]. Further formaldehyde is able to form cross links where the types A-A, G-G, A-G, A-C, and G-C were observed. In this mechanism the exocyclic -NH_2 groups of the two involved bases are coupled to form a -NH-CH_2-NH- bridge between the two. Also protein-DNA cross links can be formed in this way (textbooks on chemical carcinogens are [8a] and [8b], and in case of cross links see especially [8b] for details and further references).

Another group of carcinogens, which is a very large one, are alkylating chemicals such as dimethylsulfate $(CH_3)_2SO_4$ or methylmethanesulfonate (CH_3)-SO_2-O-CH_3. Substances like this perform a methylation, or in general alkylation, of exocyclic amino groups, nitrogen atoms in the ring systems and oxygen atoms of the nucleotide bases. For the example of guanine we show the different possibilities for alkylation (R for alkyl residue):

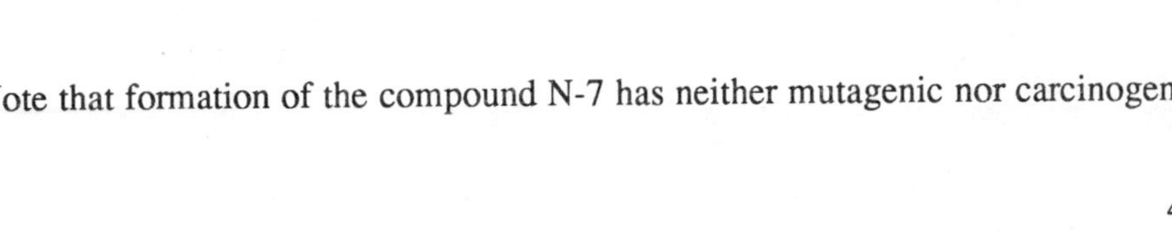

Note that formation of the compound N-7 has neither mutagenic nor carcinogenic effects. The above mentioned dialkylsulfates and alkylalkanesulfonates react mostly with ring nitrogens and are weak carcinogens, while compunds like the *N*-nitroso chemicals which alkylate mostly oxygens, including formation of phosphotriesters in the backbone of DNA, are very potent carcinogens. Examples are dialkyl nitrosamines $O{=}N{-}NR_2$, *N*-nitrosoureas $O{=}N{-}NR{-}CO{-}NH_2$ and *N*-alkyl-*N*'-nitro-*N*-nitrosoguanidine $O{=}N{-}NR{-}(C{=}NH){-}NH{-}NO_2$. The nitrosourea and *N*-alkyl-*N*'-nitro-*N*-nitrosoguanidines form directly alkyldiazonium ions $R{-}N_2^+$ which are the final alkylating agents. However, this reaction is catalyzed by hydroxyl ions and thus is not a metabolic process. Usually they act at the spot where they are administered, since they are too reactive to be transported to any significant amount. An interesting case is the activation of the Ha-*ras* proto-oncogene, since here the mechanism is quite well known. In case of activation by *N*-methyl-*N*-nitrosourea the oxygen atom of guanine is methylated, leading to the so-called O^6meG. In the replication process this leads rather easily to mispairing resulting in an O^6meG-T base pair, because one of the three possibilities of hydrogen bond formation in normal guanine is blocked due to the methylation:

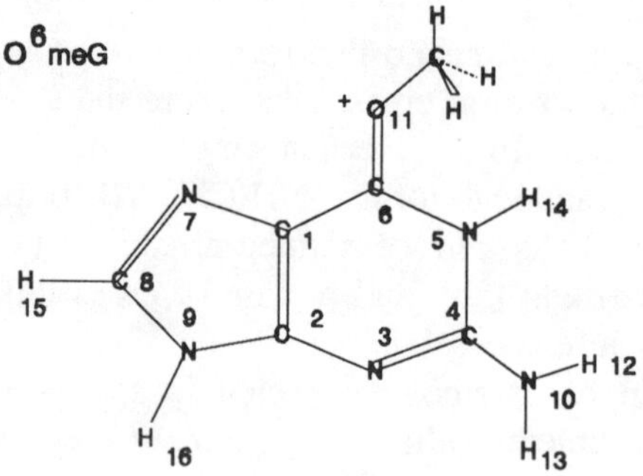

The oncogene activation occurs then in the sister generation of this mispair, where for thymine (T) instead of the guanine, correct at that site, adenine (A) is built into the sequence [9]. The methylation occurs very specifically exactly at that guanine which has to be changed to adenine for oncogene activation. The origin of this very site specific methylation is twofold: (i) special sites in DNA can be more easily attacked than others [10-12] and (ii) some special sequences cannot be reached easily by repair enzymes [13].

The *N*-nitrosamines in contrast to the other up to now mentioned nitroso-carcinogens have to be metabolically activated, however, in a single enzymatic step:

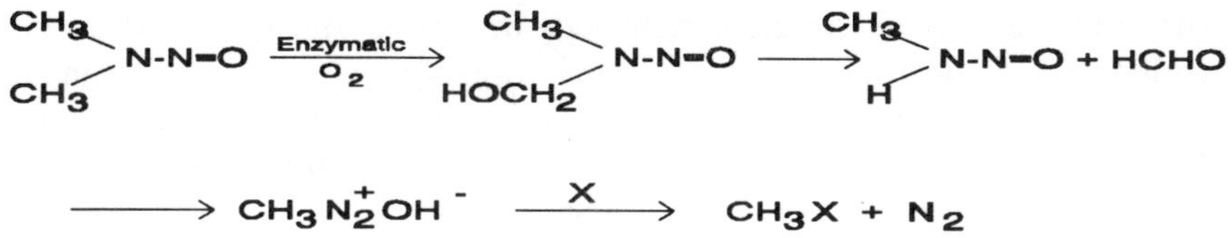

$$(CH_3)(CH_3)N\text{-}N{=}O \xrightarrow[O_2]{\text{Enzymatic}} (CH_3)(HOCH_2)N\text{-}N{=}O \longrightarrow (CH_3)(H)N\text{-}N{=}O + HCHO$$

$$\longrightarrow CH_3N_2^{+}OH^{-} \xrightarrow{X} CH_3X + N_2$$

The final alkylating agent is then again the diazonium ion. Further the nitrosamines are synthesized also in vivo (in mammals) from nitrite and from higher amines [14-17]. Another carcinogen which forms a diazonium ion as alkylating agent is 1,2-dimethyl-hydrazine (CH_3)-NH-NH-CH_3.

Also the more simple alkylhalogenides like methyliodide or ethyliodide are carcinogens due to their alkylating power [17]. However, in this case charged alkylated derivatives of the nucleotide bases are formed. One mechanism [18] for ethylbromide involves also a metabolic enzymatic initial step. The enzyme glutathiol *S*-transferase catalyzes the reaction to *S*-(2-bromethyl)-glutathiol which reacts then with guanine to *S*-[2-(N^7-guanyl)-ethyl]glutathiol. This final product in a single DNA strand tends then to mispairing with thymine and consequently, in the sister generation, guanine is substituted again by adenine.

A class of chemicals, the halonitrosoureas are bifunctional alkylating agents. The interesting point is that these compounds at a high cytotoxic level have a chemotherapeutic effect which originates from their cross linking reactions with nucleotide bases, while other reactions of the same compound lead to chain breaking and carcinogenicity. One example of such a chemical is bis(chloroethyl)-nitrosourea Cl-$(CH_2)_2$-N(NO)-CO-NH-$(CH_2)_2$-Cl. Some examples of its reaction products with nucleotide bases (besides cross links) are [8b]

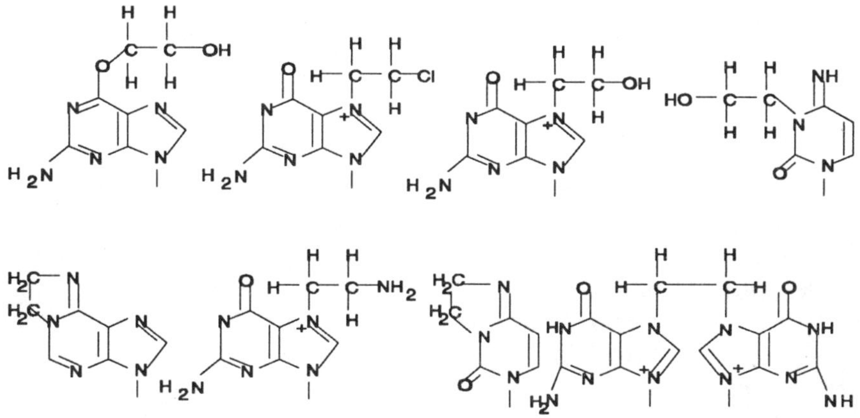

Another class of alkylating carcinogens are the cyclic ones which contain unstable and thus very reactive ring systems. Some of them are also bifunctional and thus are able to form cross links. We do not want to go into detail here but will just mention some prominent carcinogens from this family. One of them is the sulfur mustard gas:

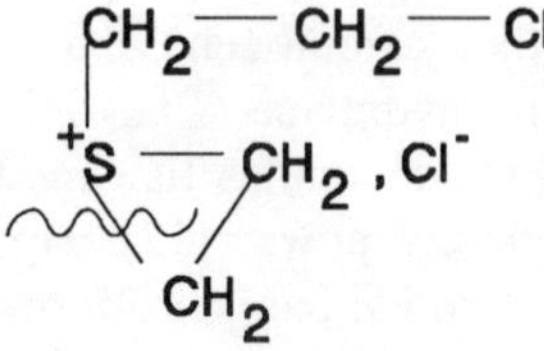

CH_2 — CH_2 — Cl
^+S — CH_2 , Cl^-
CH_2

It reacts similar to the alkylsulfates discussed above, however, in addition it is able to form cross links. A similar behavior is shown by phosphoramide mustard which results from metabolism of the chemotherapeutic and cancerogenic cyclophosphamide. A well known but not very potent carcinogen is ß-propiolactone

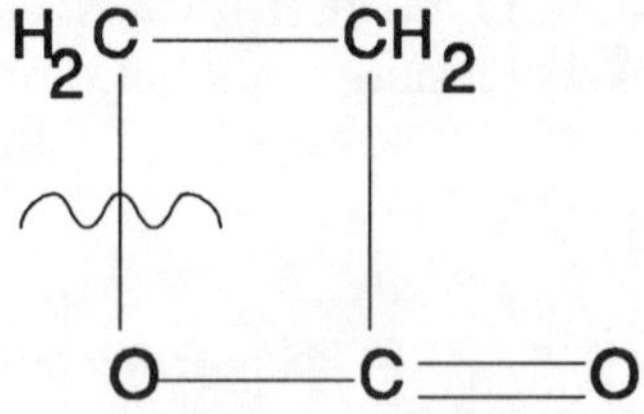

H_2C — CH_2
O — C = O

which forms carboxethyl derivatives at the N-atoms of the nucleotide bases. Also cross linking activity between proteins and DNA was observed, as well as formation of phospho-diesters at the backbone of DNA. Important carcinogens are epoxides which occur also as metabolic products of larger carcinogens as will be discussed below. An epoxide contains a three membered ring of two carbon and one oxygen atom. This ring is highly reactive and tends to open to form carbenium ions. These ions are strongly alkylating agents which mainly alkylate N7 of guanine and the nitrogen atoms of adenine leading to hydroxy-alkyl derivatives. An O^6 alkylation of guanine was not observed. The last carcinogen to be discussed at this point is vinyl chloride which has to be metabolically activated in one enzymatic step leading to chloroethylene oxide which is again an epoxide. Afterwards a ring opening occurs non-enzymatically to chloroacetaldehyde. The latter compound reacts primarily at the imino site of guanine or cytosine, leading under chloride elimination to ring formation. The resulting ring system is dehydrated to the final product in acidic medium:

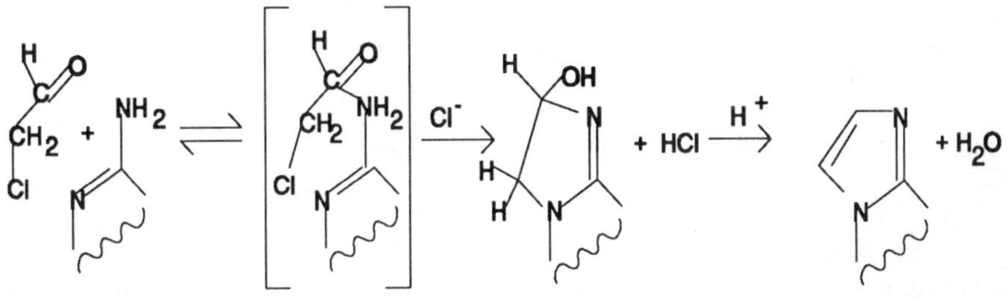

The products of such reactions which cytosine, adenine and guanine found in DNA and RNA of rats exposed to vinyl chloride are:

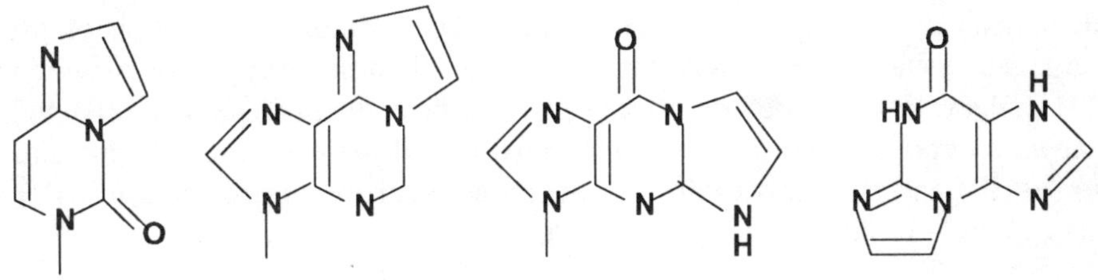

There are numerous other examples of relatively simple alkylating carcinogens which either act directly on DNA or require only one single enzymatic reaction step for their metabolic activation. However, we feel that the examples given are sufficient for the scope of this chapter and refer the interested reader to [8b,19,20] for further examples and references.

Now we want to turn to more complicated carcinogens which require several enzymatic reaction steps and rather large changes in their chemical structures for their metabolic activation. For the scope of this book we do not want to go into the details of the enzymatic system which is required in the course of such activations (see Chapter 5 of [8b] for details), but concentrate on the carcinogens themselves and on their activated forms, called ultimates and their reactions with DNA. The group of carcinogens of this kind we want to discuss first are the polycyclic aromatic hydrocarbons, most of which e.g. are present in cigarette smoke. Some examples of the numerous polycyclic aromatic hydrocarbons are, from left to right, phenanthrene, benzo[a]pyrene and benzo[e]pyrene:

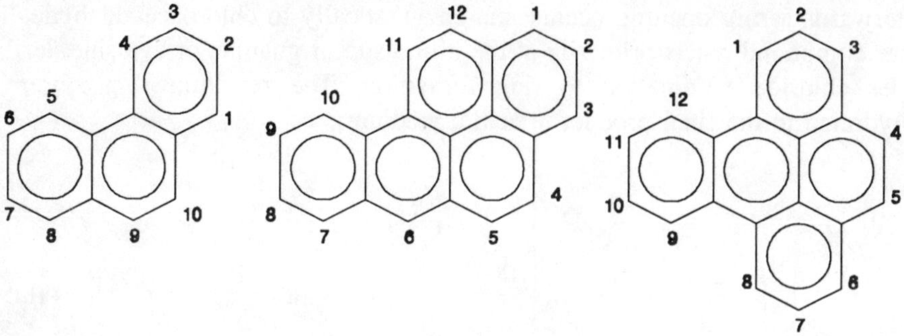

The compounds benzo[a]pyrene and 2- and 3-acetaminofluorene (2-AAF, 3-AAF), where the latter two are not pure PAH's, are the most widely studied examples [21-23] in this large family of potent carcinogens [24-26]. The formation of the adduct of the ultimate of benzo[a]pyrene with guanine by electrophilic substitution at the 2-amino group of guanine was even studied theoretically in model calculations using quantum mechanics [27]. The metabolic activation of these compounds usually follows a common pathway. By a monoxygenase enzyme the PAH (polycyclic aromatic hydrocarbon, I) is oxidized to an arene oxide (II) which is again an epoxide. With the enzyme epoxide hydrolase a *trans*-dihydro-diol is formed (III) and from this again with the monoxygenase enzyme an *anti*- (IV) or *syn*- (V) dihydrodiol epoxide is obtained:

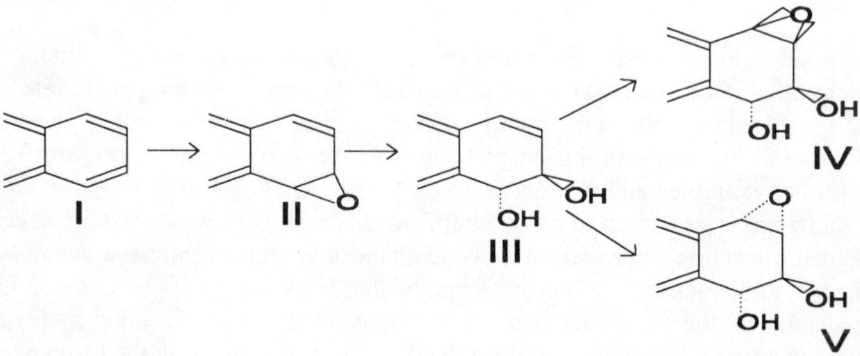

The resulting diol epoxides can now either be detoxified to a tetrol or, as an electrophilic agent, bind to nucleophilic sites of the nucleotide bases. The mechanism of the possible detoxification will not be discussed here. It is interesting that there were attempts to relate the stability of the carbenium ions formed by ring opening of the diol epoxides to the carcinogenic activity of the parent PAH's [28]. However, only a partial correlation was found [29,30]. The formation of the main (there are isomers formed of the same chemical composition but with different relative orientations of the substituents) ultimate of benzo[a]pyrene (BP) starts with the formation of (+)-BP-7,8-oxide by a mixed function oxidase system. From this the enzyme epoxide hydrolase builds the (-)-BP-7,8-dihydrodiol and finally the ultimate (+)-BP-7,8-diol-9,10-epoxide is synthesized again with the mixed function oxidase system [31-34]:

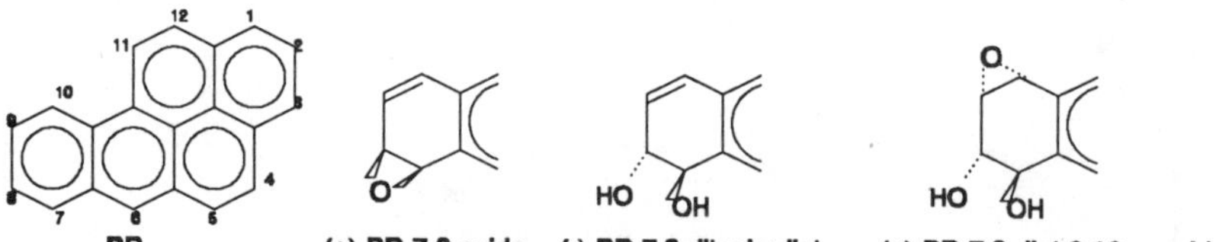

BP (+)-BP-7,8-oxide (-)-BP-7,8-dihydrodiol (+)-BP-7,8-diol-9,10-epoxide

The formation of the diol epoxides of the carcinogenous PAH's uniquely occurs in a saturated angular benzene ring and always at a part close to the so-called bay region of the PAH's, which is a sterically hindered region in them. Examples are the bay regions of phenathren (I) and benzo[a]pyrene (II):

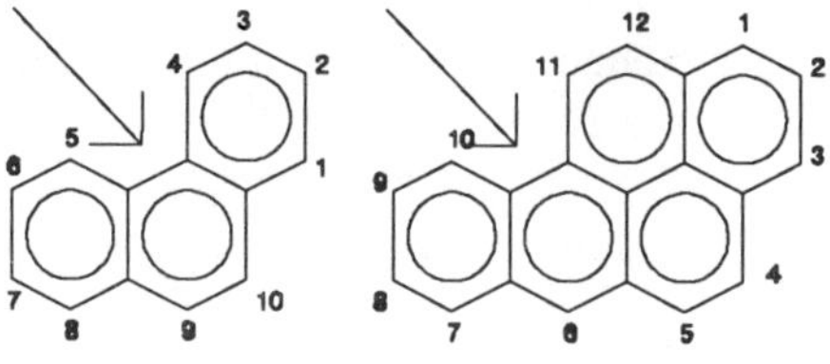

I II

The epoxides formed at such a sterically hindered bay region are supposed to be highly reactive and to undergo ring opening for formation of the alkylating carbenium ion more easily than epoxides in other regions of the PAH:

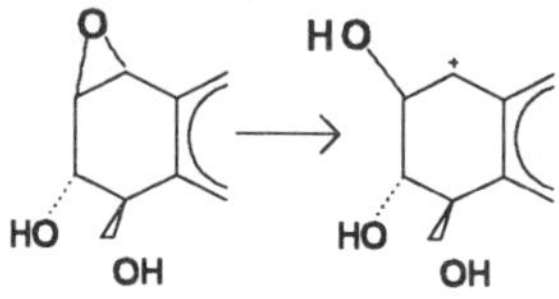

Another reason why bay region epoxides are the most powerful mutagens and carcinogens is quite probably that the sterical hindrance caused by this region inhibits the detoxification of the epoxides to tetrols by the epoxide hydrolase enzyme.

After the ring opening of the epoxides the resulting carbenium ions are able to alkylate nucleotide bases, especially guanine. The adducts resulting from alkylation of guanine by the ultimates of benzo[a]pyrene (a) and 2-AAF (b) are [35a]:

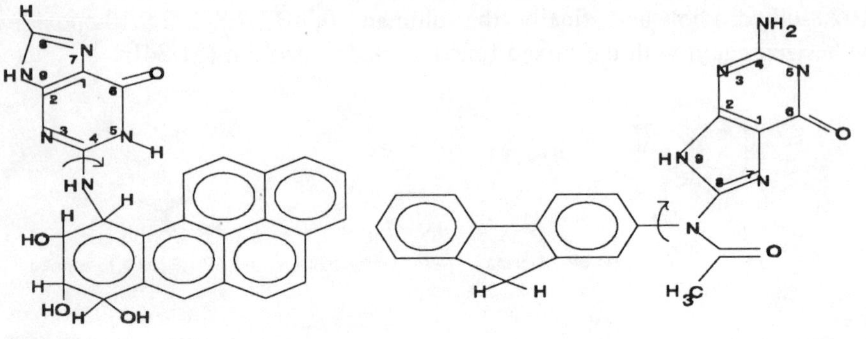

(a) (b)

The mechanism of metabolic activation of AAF will be discussed briefly below. Now the question arises, what causes the mutation after the alkylation of guanine and other nucleotide bases in a DNA strand with the very bulky ultimates of the PAH's. It is known from experimental evidence (see [8b]) that the ultimate of benzo[a]pyrene bound to DNA does not intercalate into DNA. Further from model building it seems that also the conformation of the DNA molecule is not changed too much. The conclusion is that the carcinogenic activity of the guanine bound to this ultimate relies on the change in hydrogen bonding between the modified nucleotide base and its counterpart on the second strand. However, details are not well known yet (see [8b] and references therein). It is known from theoretical calculations [35b] that the charge distributions in the two parts of this system are considerably changed resulting in the transfer of negative charge (electrons) from the benzo[a]pyrene ultimate to the guanine residue. Such a charge transfer should affect considerably the properties of guanine concerning hydrogen bonding. Thus mispairing might occur when this part of DNA is expressed. Another possible mechanism, namely soliton creation from a geometrical distortion will be discussed in detail in the later chapters of this book.

A completely different case is the above shown adduct of the ultimate of 2-AAF (and also that of 3-AAF) to guanine in the C8 position. It is well known that planar aromatic compounds which are not covalently bound to the nucleotide bases can intercalate. That means that they form complexes with DNA in which they reside between two nucleotide base pairs of the macromolecule. In the above shown adducts a simple rotation around the glycosil bond turns the guanine residue outside the helix, the so-called base displacement and the ultimates of the AAF's inside, in an intercalation-like manner:

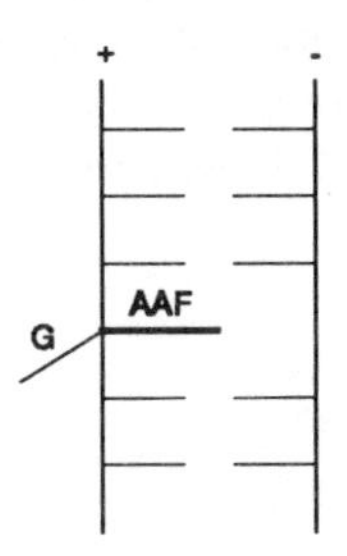

Because such a rotation occurs most easily in a single strand it is assumed that it actually takes place in single stranded DNA's. Since AAF cannot form hydrogen bonds the presence of this molecule in DNA causes a base deletion, a so-called frameshift mutation. This mechanism has been suggested by Slater and Riley [36]. The insertion of AAF causes an enlargement of the distance between the neighboring two base pairs. However, experiments suggest that the distortion is not that much restricted to two sites, but extends over about 5-50 sites [8b]. A more accurate determination of this extension was not possible. Such a structural change will not change the stacking interactions too much because they are anyway small [37-39] and may be counterbalanced by interactions between AAF and the nucleotide bases [35b]. It was confirmed that in vitro as well as in vivo the same products are formed in the reaction between DNA and the ultimates of the AAF's [40]. In case of alkylation of the *N*2 amino group of guanine which happens in 20% of all cases the above rotation does not take place [8b,41]. However, the mutagenic event is certainly the base displacement process [42].

The above discussed AAF falls into the general class of carcinogens called *N*-substituted aromatic compounds which are sometimes also termed as aromatic amines, although not all of them are amines. Some examples are (a) 4-aminobiphenyl (b) benzidine and (c) 2-naphthylamine:

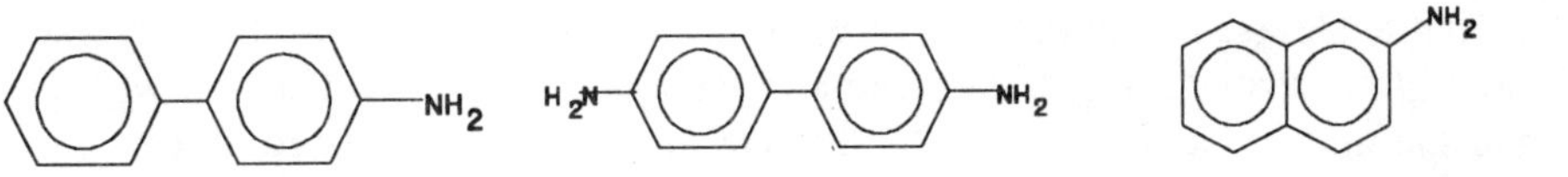

NH_2 H_2N NH_2 NH_2

a b c

Potent carcinogens of this kind mostly contain two or more aromatic rings. Simple aromatic amines like aniline or acetanilide are not carcinogenic and e.g. *o*-toluidine is considered as a weak carcinogen. All these carcinogens require metabolic activation to exhibit carcinogenic activity.

The different molecules of this kind are usually metabolized in different ways. As a representative example we will here describe the metabolization of AAF [43,44]. To become active the molecule has to be transformed again into an electrophilic agent to be able to bind to DNA. In a first activation step AAF is oxidized by the cytochrome P-450-dependent monooxygenase enzyme system in the liver to *N*-hydroxy-AAF:

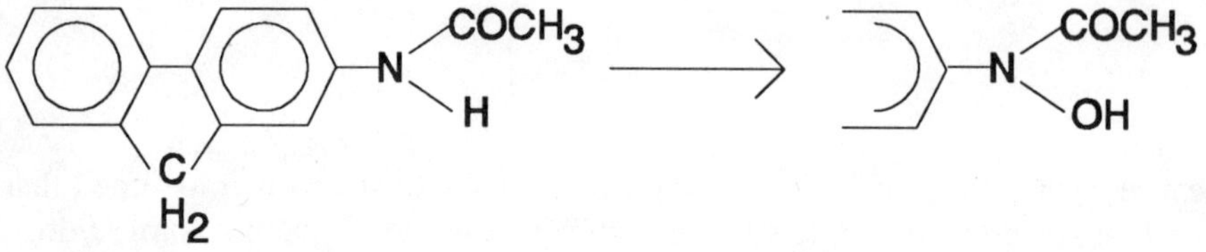

Further with the help of a soluble sulfotransferase enzyme from the *N*-oxidized AAF the very reactive AAF-*N*-sulfate ester is synthesized:

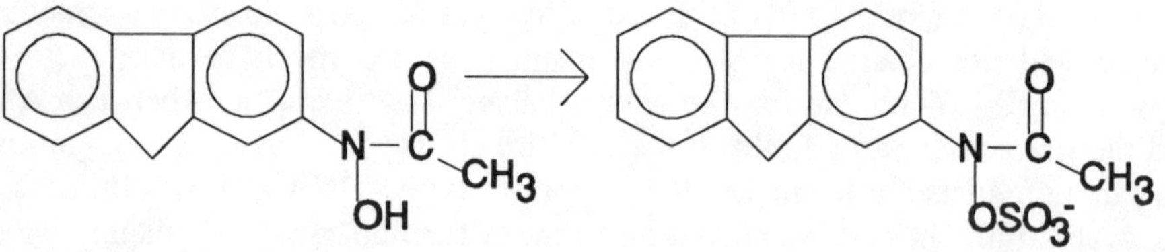

This highly reactive molecule according to all evidence seems to be the ultimate carcinogen of AAF. The sulfate group is very easily removed and the remaining cation can attack DNA to form the above described products. However, note that this is not the only metabolic pathway by which the *N*-hydroxy-AAF can be converted into the ultimate carcinogen. Other enzymes like e.g. *N,O*-acetyltransferase and deacetylase are also capable to form the ultimate carcinogen from *N*-hydroxy-AAF (for more details see again [8b]). The reaction with DNA and the mutagenic activity of the ultimate of AAF was discussed above.

Another family of carcinogens which bears some structural relationships to the aromatic amines are the azo compounds (see Chapter 10 of [8a] for details). The azo compounds mainly induce liver tumors and are taken up with the food.

From the large family of azo dyes we only want to show the chemical structure of the 4-aminoazobenzenes:

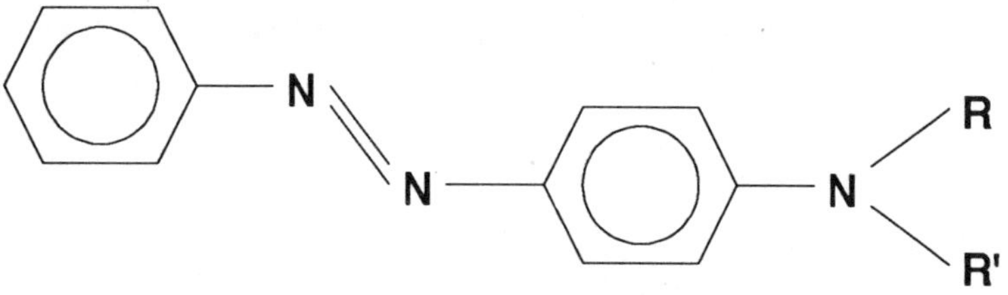

Depending on the nature of the two alkyl rests at the amino group the compounds may be carcinogenic or not. For example the well known dye butter yellow with two methyl groups at the N atom (4-dimethylaminoazobenzene or DAB) is known to be carcinogenic. The benzyl ring which does not contain the amino group can also be substituted by other aromatic residues like a naphthyl or a pyridinyl ring without loss of carcinogenic activity. As well as benzyl rings toluyl groups also produce carcinogenic compounds. We do not want to go into details of the metabolization of the azo dyes. Reductive fissions of the azo group, hydroxylations, demethylations and other reactions occur. Some examples of metabolic products of DAB after reductive fission are:

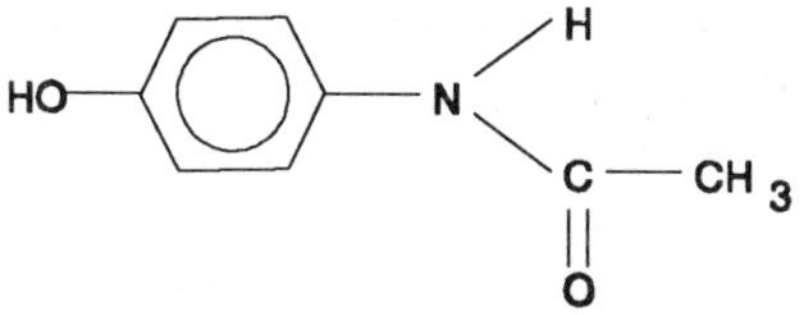

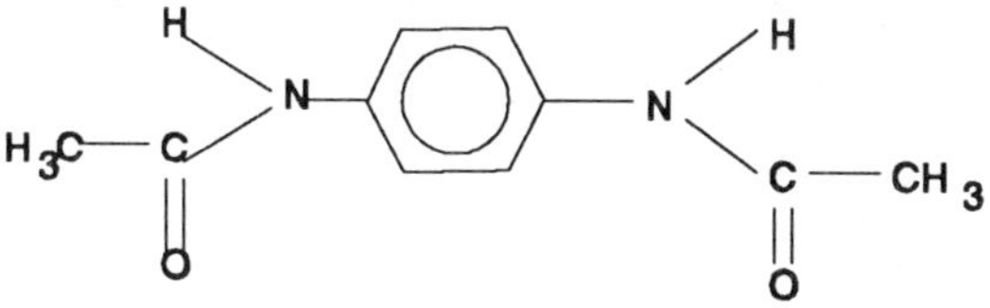

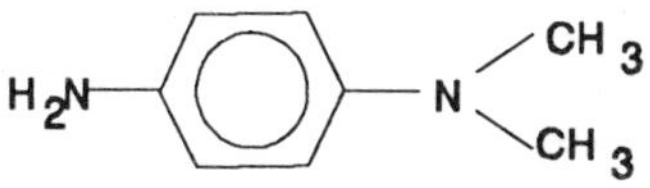

It is known that the azo dyes bind to DNA as well as to proteins. An interesting finding is that from DAB containing ^{14}C atoms the ring carbons are not incorporated into liver nucleic acids, but the methyl carbons are to a large degree.

To close this section, for the sake of completeness, we just want to discuss very briefly some naturally occurring carcinogens which are mainly found in food contaminated by fungi. The first kind of this species are the aflatoxins which were found in peanuts contaminated by Aspergillus flavus [45]:

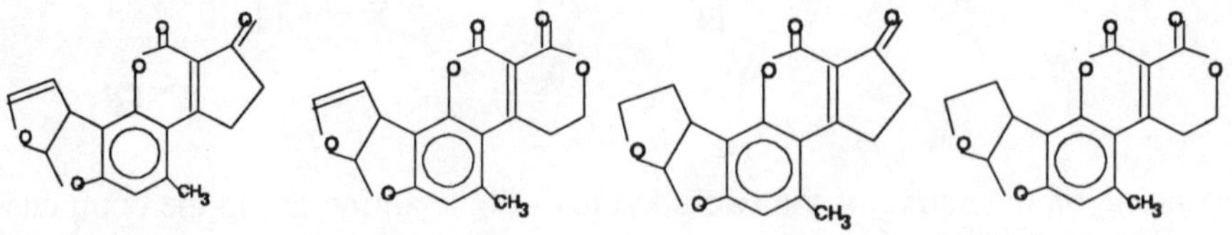

The aflatoxins are known to cause hepatomas in fish and humans, the latter mainly in Africa and South-East Asia. However, to give details of the metabolic activation process of aflatoxins seems to go beyond the scope of this book. Further naturally occurring carcinogens, like sterigmatocystin, safrole (from oil of sassafras) or estragole (1-allyl-4-methoxybenzene) we only want to mention here and refer the reader again to Ref. [8b], Chapter 6 and references therein.

In summary we can conclude that all known carcinogens, either directly acting or in metabolized form attack the nucleotide bases, sometimes also the phosphate groups or the sugar residues of DNA, causing mutations directly or at least strong disturbances in the geometry or charge distribution of DNA all of which finally can lead to the activation of proto-oncogenes in the genome of a normal cell. More details about these mechanisms will be discussed later in this book.

3.2 The Role of Electromagnetic and Particle Radiation

Besides chemical carcinogenesis, which accounts for most tumors cases [46] it is known that exposure of the body to radiation like electromagnetic ultra violet (UV), X- or γ rays or particle beams as radioactive α- (helium nuclei) or β- (electrons) radiation is a frequent cause of cancer. The question of the effects of UV light is becoming especially important nowadays because of the formation of the well known and widely discussed ozone holes in the upper part of the atmosphere, causing an increase of the UV light intensity reaching the Earth's surface, especially those with dangerous frequencies. It is known that ultraviolet light with wavelengths around 260 nm kills bacterial cells and the surviving population contains many mutants. However, these mutants are much more sensitive to UV radiation than the normal type of cells. This suggests that in normal cells a repair mechanism is present which does not function in the mutants, i.e. is destroyed by the radiation. This explains also why the number of surviving cells as function of the UV dose is not a linear one, since the number of more sensitive mutants in the surviving population increases [47]. Now we have to ask, what radiation actually does in cells to cause mutations, which finally are also the origin of the cancerous activity of radiation.

First of all radiation of sufficiently high frequency (energy) can form radicals (highly reactive chemical species containing an electron with an unpaired spin) in the cell which in turn due to their high reactivity can attack DNA and lead to mutations. Such a process might be viewed as a variant of the activation of carcinogens. We will come back to this possibility later. The most frequent chemical change in DNA caused by UV radiation is the formation of a cyclobutane ring between two neighboring thymine molecules in one strand:

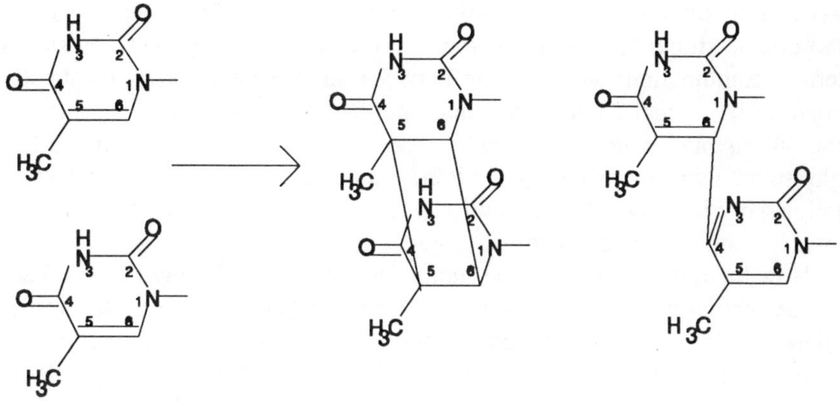

The formation of such a thymine pair obviously must have drastic consequences on the structure of the DNA segment surrounding it. Another unusual chemical bond caused by UV irradiation is shown on the right of the above sketch, the so-called TC(6-4) product which also causes strong geometrical distortions. In bacteria and in less developed eukaryotic organisms (not in mammals up to now) an enzyme called photolyase was found which is able to break these new bonds directly if it is activated by visible light [48-52]. A review on DNA repair enzymes can be found in [53]. In higher organisms, as well as in prokaryotic cells, in addition to the photolyase, a complex system of enzymes synthesized from genes called uvr (from UV repair) A to D is able to repair such DNA damage by a so-called excision mechanism. The uvr endonuclease coded by the *uvrA*, *uvrB* and *uvrC* genes cuts out from the chain the damage together with some nucleotides on both sides of it. Then the DNA-helicase enzyme coded by the *uvrD* gene removes the cut segment from its complementary chain and the gap is subsequently closed by DNA-polymerase I [53-57]. It is this main repair mechanism which is destroyed in the mutants of *Escherischia coli* discussed above.

In addition, exists a third repair mechanism exists which needs neither photoactivation like the photolyase, nor excision. It is the so-called post-replicative repair (see [58,59] for reviews and further references). Let us consider a DNA double strand containing, say two pyrimidine dimers induced by UV irradiation, each on one of the two single strands (note that exposure to UV light always leads to more than one pyrimidine dimerization). These dimers are sites where the replication of DNA stops some nucleotides before the damage and resumes again after it. Thus after the replication we obtain two double strands each of them containing again a pyrimidine dimer and one broken chain. In a subsequent replication cycle we would then obtain broken DNA segments, i.e. so-called double strand breaks (see below). The next step of the repair is similar to genetic recombination: an exchange between strands happens enzymatically. A segment of the strand without the break, but having a dimer is put into the break of the other strand without a dimer. In this way the cell finally obtains a broken single strand, two strands with pyrimidine dimers and one completely intact single strand, from which by replication an intact double strand is obtained and the full genetic information is regained.

Now the question is how the mutagenic activity of UV light arises, because the three repair mechanisms described restore the DNA to its state before the irradiation. In connection with this question, an important reaction of cells against any damage of DNA including that arising from UV exposure, the so-called SOS system was discovered, in particular by Weigle [60] and Witkin [61,62]. For details and further references see the review articles [58,59]. The name SOS comes from the fact that the cell in this case obviously sends a signal which causes all available enzymes for DNA repair to become active also those with a rather large error rate in their repair activity. This causes reading errors and thus mutations. This is called error prone repair. It is known that the so-called SOS system consists of at least 12 genes which are activated in the case of strong DNA damage. The SOS reaction is activated by the product of the so-called *rec A* gene. This gene is activated by single stranded DNA. The rec A protein is a protease and cleaves predominantly the so-called lex A protein which is the repressor for the gene sequences coding the SOS system. After cleavage of the lex A protein it leaves its binding place at the DNA and the SOS genes are free for expression. The rec A protein is again deactivated when the SOS system had done its repair job and no more activating single stranded DNA is available. Subsequently lex A is synthesized again and blocks the SOS genes. For details of this very complicated process see the reviews [58,59]. To mention some examples, in the course of the SOS reaction of the cell also the *uvrA*, *uvrB* and *uvrC* genes become higher activated. Other products of the SOS system stop the cell division or block the activity of DNA-decomposing enzymes.

For our purpose the error prone repair genes which are also activated in the SOS system are of interest because they cause mutations due to erronous repair. In this respect specifically the so-called *umu C* and *umu G* genes (umu from UV mutagenesis) are relevant. It is assumed that the products of these genes serve to fix the DNA polymerase III to DNA at sites where the strand geometry is significantly disturbed (e.g. at pyrimidine base pairs). From sites of this kind, as mentioned above, the polymerase usually decouples. With the help of the umu proteins it stays at the DNA and continues to polymerize nucleotides, but, due to the distorted matrix, with a large probability of errors. However, this part of the mechanism is not fully clear and there are other possibilities under discussion (see e.g. [47]). In this way UV radiation can induce mutations primarily [62-64] via formation of pyrimidine base dimers and the cell's own repair activity. Thus finally through the mutations the activation of oncogenes is possible. There are some other aspects of UV induced mutagenesis mostly in special cases under discussion [65-69], however, their description would go beyond the scope of this book. Since this Chapter is concerned with a biochemical discussion of mechanisms we will postpone the description of soliton creation after pyrimidine dimerization or nucleotide base excitation until a later part of the book. Besides the pyrimidine dimers (thymine and cytosine) [70,71] also purine-pyrimidine dimers have been found [72]. A review is given in Ref. [73]. Finally one should mention that damage due to UV light preferably occurs at special sequence parts the so-called hot spots of DNA.

An interesting example of carcinogens are the so-called psoralens. Like the polycyclic aromatic hydrocarbons they posses a large planar π-electron system. Also like the PAH's, they need activation to be carcinogenic. The psoralens themselves just intercalate into DNA and do not lead to damage. Interestingly, their activation occurs not via metabolization as in case of the PAH's but by UV irradiation (wave length 300-380 nm). Only after irradiation can they bind covalently to DNA and besides other adducts they are able to form cross links between bases which are believed to be biologically the most important reaction products [74-76].

The next point in this connection is the exposure of cells to ionizing radiation, i.e. electromagnetic radiation of higher energy (frequency) like γ- and X-rays or particle radiation. These forms of radiations interact with the electrons of the material they hit and in this way lose energy. This lost energy is usually enough to ionize other molecules. More theoretical details about the processes involved, like plasmon (collective electronic excitations in solids) formation and decay we must postpone to later chapters of the book. The action of such radiation on biological macromolecules like DNA can involve a direct interaction with the macromolecule or first of all interactions with molecules in the cell, predominantly water. In the latter case electrons and chemical radicals are set free and interact subsequently with the biologically important molecules. The extent of the damage done by such radiation in the cell naturally depends on the amount of radiation hitting the cell, i.e. on their dosage, and secondly on the actual kind of radiation. γ- or X-rays usually do not cause too much ionization and thus can penetrate deeply into the material, while particle radiation causes a high ionization density, but does not penetrate very deeply into the tissue. The proliferation of energy into a tissue is usually expressed by the dose, which is defined as the amount of energy proliferated divided by the mass of the irriadiated tissue, i.e. in Joule (J) per kilogram (kg). The unit for this dose is 1 Gy (Gy for "Gray") where 1 Gy = 1 J/kg. An older dose unit, which is still more common, is the rad, which is based on an older energy unit, the erg. Here 1 rad = 100 erg/g and thus 1 Gy = 100 rad. Since the damage caused by a given dose in a given tissue is larger for particle radiations than for γ- or X-rays, a further unit was introduced, to take this factor into account. In the older unit system it was the rem, in the new system the Sievert (Sv). Roughly speaking 1 rem is the dose of any radiation which correponds to the biological effect of 1 rad γ-radiation and 1 Sv = 100 rem. To convert rad into rem commonly a factor of 1 is used for X-rays and a factor of 10 for α-rays (helium nuclei). Thus X-rays cause as much damage in a tissue as γ-rays, while α-radiation is able to cause 10 times more damage than the same dose of γ-radiation. According to Upton [77] the dose per year an average person in Western Europe or North-America is exposed to from natural and civilization related sources is 1.8 mSv/year = 180 m rem/year. According to [78] a full body irradiation of a person causes no damage for a dose between 1 and 2 Sv. For 2-4 Sv a lethality of 10-25% shows up. This increases to 50% mortality for 4-6 Sv and up to 100% for doses larger than 6 Sv. Naturally, in the course of irradiation, the DNA is damaged. However, for higher dosages this is unimportant, because the patient dies from other damage before mutation can play a role. For smaller dosages (less than 5 Sv), on the other hand, this is very important, since in these cases mutations caused by the irradiation lead to development of cancer, especially leukemia in the survivors. This became very clear in investigations on the survivors of the atomic bomb detonations at Hiroshima and Nagasaki.

The DNA damage which is easiest to detect is the so-called double strand breaks (DSB's, see below for more details) and the following end-to-end reunions which lead to abnormal chromosome forms (see e.g. [79]). However, there are many other kinds of damage known which are caused by ionizing radiation either due to direct interaction of the radiation with DNA or by indirect effects. Those are the formation of cross links within a base pair, cross links between different chromatids or breaks of a complete single strand of the double helix (single strand breaks, SSB's, also called chromatid aberrations) which are known to occur very frequently but, as we know from above, can, to a large extent, be repaired, e.g. by the SOS system. However, as previously discussed, especially the activation of the SOS system can lead to misrepairs and consequently to cancer. Furthermore the whole double strand can be broken (double strand breaks, DSB's), i.e. both strands of a double helix can break at nearby sites. Last but not least, the structures of the individual nucleotide bases can also be drastically changed, and it is even possible that the whole base is deleted from the chain (see e.g. [8b]). For thymine alone, 24 different irradiation products are known. As mentioned before, some of this damage can be repaired by the uvr enzyme system, some others cause the activation of the SOS system, with all the consequences discussed above. These are a consequence of irradiation with doses up to 5 Sv.

An important question is now if lower dosages, like those which occur after X-ray checkups or e.g. due to the reactor catastrophe in Chernobyl, can lead to carcinogenesis or other mutations. The basic problem is whether the repair mechanisms of a cell are efficient enough to repair the effects of a small (below some fixed threshold value) dose of radiation completely, or if any dose of radiation can eventually lead to mutations. Answers to these questions are mainly known from experiments on animals. Concerning mutations it was found that after irradiation of the sperm cells of monkeys with a dose of 0.01 Gy, 1 out of 1,000,000 new born animals will have damage due to mutations. Further it was found that the mutation rate caused by a given dose of radiation is much larger if the amount of radiation is given in several high doses than if the same amount is given during a longer period of time. From this one can conclude that in the latter case the repair systems of the cell are able to counteract the damage [81]. Experiments with cell cultures showed that even an irradiation of 1 Gy doubles the natural mutation rate. Also, other findings indicate that any dose of radiation increases the mutation rate and thus no threshold dose exists below which no mutations are caused [82].

In the remaining part of this section we want to turn shortly to the questions concerning the kind of mechanism which might lead to the observed chromatid and chromosome aberrations in cells after irradiation. A chromatid aberration is essentially a single strand break, while a chromosome aberration requires a double strand break. It was found that whatever type of break occurs it can also be repaired and thus leads to a reduction in the frequency of observed aberrations. Further, chromosome aberrations occur more frequently prior to DNA synthesis, while chromatid aberrations during or after the replication process [83,84] (for a more recent review see [85]). Later several workers demonstrated that the kinetics, i.e. the dose response Y (number of aberrations per time and cell), for X-ray induced aberrations have a linear and a quadratic term in the dose D [86-89]:

$$Y = a + bD + cD^2$$

Here a is a background frequency, and b and c are the linear and the quadratic coefficients, respectively. This result is important, because the kinetics show that any kind of aberration is produced either by one or by two subsequent interactions of the radiation with DNA, depending on the dose.

Nowadays (see [106] for references), the above kinetics are more exactly measured leading to [89]

$$Y = (a + bD + cD^2) \exp[- (dD + eD^2)]$$

The exponential decay occurs, because for higher dosages the cells simply die before any mutagenic or carcinogenic effects can occur. There are basically two concepts to explain the part of the kinetics dominating at lower dosages. The first and today generally accepted one [85] is the so-called breakage first hypothesis [83,90]. In this hypothesis, single breaks in the strands caused by radiation are the primary events leading to aberrations. These broken chromatids can afterwards be rejoined normally by the repair systems of the cells, they can be left unrepaired or can participate in exchange mechanisms as described above for the UV case. Later [79,91-93], the so-called exchange theory was worked out. In this hypothesis it was assumed that the radiation does not initiate a break, but an exchange of parts of chromatids. If this exchange happens in an incomplete way it should give rise to discontinuities in chromatids and subsequently to breaks, which would lead to slightly different kinetics than the breakage first hypothesis.

These two hypotheses were a matter of intense debate for some years (see e.g. [94]). Currently the breakage first hypothesis is the most widely accepted one, although it is still assumed that some proportion of aberration might also be due to the exchange mechanism [85] (see also [94-96] for details). From the results discussed, it is obvious that the production of a chromosome aberration requires a double strand break, i.e. the break of both chromatids of a double stranded DNA where the breaks have to be located close to each other. This might happen by a single interaction (one track) of radiation with DNA or by two consecutive (two track) ones.

However, especially for low dosages such processes seem to be rather rare events. This problem which might be solved by the concept of soliton formation is discussed in more detail in later chapters. How the chromosome aberrations are formed in the case of low radiation doses is still a matter of debate. They might occur via errors during direct repair of the DSB's or maybe also in misrepairs of damage done to bases during the previously described excision repair mechanism (see [97-99] for details). It is improbable that direct ligation of DSB's occurs and nowadays a recombination-like mechanism is discussed as the most likely one for aberration formation, which is also assumed to play the major role in the case of higher dosages of radiation where directly induced DSB's become probable. Finally we want to mention, postponing a more detailed discussion again to the later chapters of this book, that a statistical theory for the probability of double strand breaks in DNA has been worked out [100]. Also first steps have been taken towards a quantum mechanical calculation of the probability for the primary events of DSB formation [101]. However, the links between the two different types of theories are not yet established [102].

3.3 Other Factors Involved in Cancer Initiation, e.g. Role of Trace Elements in the Cell, Heat or Continous Local Pressure

It is also known that the lack or abundance of special chemical elements which are usually either not present at all or only in traces in the cell can influence carcinogenesis. An example is selenium, where laboratory and epidemiological studies have led to the conclusion that presence of this element can offer restricted protection against the risk of cancer. It is active in the form of selenites or selenates both in an anti-initiator and anti-promotional way [103,104]. The detailed mechanism for that is unknown. However, it is known that selenium protects cells from oxidative damage. It is an important part of the enzyme glutathione peroxidase which is necessary to catalyze the reduction of lipid peroxides and of hydrogen peroxide. Treatment with selenium doubles the cell's ability to destroy such peroxides. Similar studies on other minerals such as iron, copper, zinc, molybdenum, iodine, arsenic, cadmium, and lead if present in food have led to no conclusive results [104]. Molybdate and sodium have been found to be effective against tumors in rats [103]. Also here, the mechanism is unknown.

The role of the brain and of the central nervous system in carcinogenesis is discussed in Chapter 7 of this book. Certainly radioactive substances are carcinogens since they emit radiation. Interestingly, even continous exposure to heat is found to be carcinogenic, causing mainly squamous-cell cancer in various mammals [105]. Furthermore, it is known that the continous pressure of a needle on a tailor's thumb, of a hammer on the hand of a shoemaker or of the bridle in the mouth of a horse can lead to cancer after years [105]. Also in this case no mechanism of carcinogenesis is known. In addition, other physical carcinogens exist such as dust particles from concrete, rubber particles from tires or asbestos fibres when inhaled. A recent review on the effects of ionizing radiation as a physical carcinogen can be found in the book edited by Weisburger [106].

The last point we want to mention here is that persons exposed to extremely low frequency electromagnetic fields [107] at work and from power lines near their houses [108] show increased cancer rates. The same holds for magnetic fields for special cancer types [108]. There is not much known about the mechanisms of the carcinogenesis due to magnetic fields, however, the evidence indicates that magnetic fields do not show tumor-initiating activity, but tumor promoting activity. Furthermore, it seems that there is a window of magnetic field strength within which the fields become active, while above or below this window they do not seem to have carcinogenic effects [108].

3.4 The Role of a Second Group of Chemicals: Cancer Promoters

Tumor promoters are chemicals which do not initiate cancer by themselves, i.e. they are not carcinogens in the strict sense although some of them are known to create a very low number of benign tumors themselves after repeated application. However, if they are applied repeatedly after the action of a normal carcinogen, they accelerate the rate of tumor development. Thus promoters can also lead to cancer development if only a low dose of a carcinogen, called the initiator is applied, which alone would not lead to cancer. These concepts of initiation and promotion were discovered in 1944 [110]. In this sub-chapter we want first to discuss briefly the biological and chemical effects of promoters, then describe the chemical structures of some representative promoting molecules. Finally we will give a short description where in the cell these molecules bind and what their probable activities are there.

The concept of two-step carcinogenesis in mouse skin was developed and further worked out after 1944 [111-113]. In this work the cells were treated with a low amount of a carcinogen, namely benzo[a]pyrene which we discussed above. This amount does not lead to carcinogenesis, but it activates some normal cells which are, however, dormant. This first step in carcinogenesis is irreversible and called the initiation. Afterwards croton oil was applied repeatedly. Later the phorbol diester 12-*O*-tetradecanoylphorbol-13-acetate (TPA) was identified as the most active promoter in croton oil [114,115]. If the promoter treatment is carried out long enough carcinomas develop. However, this step is reversible, i.e. if the application of the promoter is stopped before development of carcinomas no carcinogenesis is observed. It was also found that promoters like TPA alone cause large biological effects on the skin which, however, are reversible. The promotion in mouse cells can be divided into two distinct stages, which are both promoted by TPA. However, there are selective promoters known, which promote only one of the two stages. In stage I, the initiated cell is converted into a dormant or latent tumor cell. Thus this stage is called conversion. In stage II, the propagation of these cells is promoted [116-118]. Several chemicals have been found which act selectively as promoters or inhibitors of either stage I or stage II of promotion. Further it was found that initiation and two-stage promotion is still not sufficient to cause formation of carcinomas in mouse skin, but that there are at least two mutagenic steps involved. From this it was concluded that the role of promotion is probably to increase the number of cells which carry one of the necessary mutations for carcinogenesis and in this way to increase the probability for the occurrence of the second mutation necessary for malignancy [119-121]. This finding complements the previously discussed necessity of interaction of different activated oncogenes for carcinogenesis. However, we cannot go further into the details of these complicated biological and biochemical mechanisms. Besides promotion of carcinogenesis in mouse skin also numerous other kinds of tissues, organs and cells are known where tumor promotion has been proven experimentally. Here we only want to mention one example, that is the promotion of rat liver hepatocarcinogenesis by phenobarbital [122,123]. References on different types of systems where promotion was found are e.g. [124-127]. Let us now turn to the chemical structure of some representative tumor promoters.

The first identified (see above) and widely studied tumor promoter is the phorbol diester TPA. The phorbol, where the ester residues are represented by R^1 and R^2 is [128] (from Ref. [125])

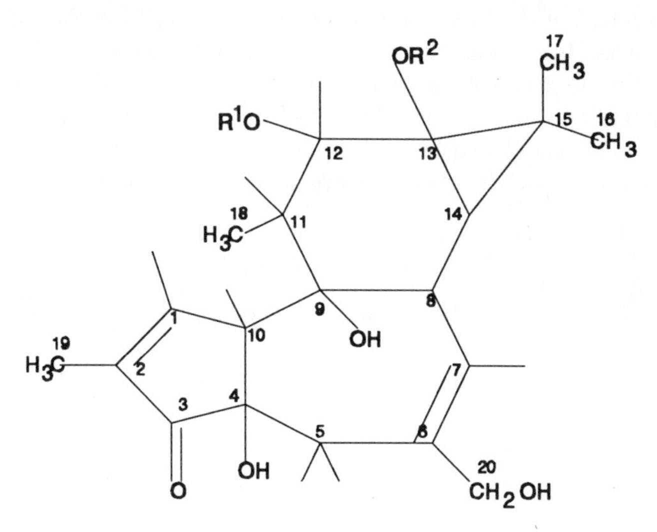

Several derivatives of phorbol diesters with fatty acids (terpenes) have been synthesized and checked for promoting activity. Phorbol itself is inactive [129]. Although the phorbol diesters are not steroids, they show some similarities to them in their chemical structure (for a short discussion of steroids see below). The compounds with the highest promoting ability are those with a long chain ester group either in position C12, called A Group or in position C13, called B group. The highest activity is found for TPA as discussed above. More detailed investigation showed that the biological activity of phorbol diesters require either a fatty acid ester at C13 or a combination of two fatty acid chains at C12 and C13 with a length of together 14 to 20 carbon atoms. For a review see [124,128]. Further one finds that the parts of the phorbol system itself which are important for promoting activity are the allylic hydroxyl (-OH) group at position C20 in free form (e.g. not alkylated) and the steric *trans* configuration at C4 and C10 where the two rings are joined. The *cis* configuration is inactive. The hydroxyl group at C4 can be lost without inactivation, however, a compound where the oxygen at C4 is methylated is only a weak stage I promoter [130,131]. Further it is necessary for promoting activity that the molecule contains a hydrophilic region together with a hydrophobic one. From different plants several kinds of diterpene esters have been isolated which exhibit promotion activity. We cannot dicuss them here in detail, but rather refer the interested reader to the review by Hecker [128]. However, one should mention that this large variety of diterpene esters should help in elucidating the mechanism of tumor promotion.

The first member of another series of skin tumor promoters was first isolated from *Streptomyces mediocidicus* and called teleocidin. It is an indole alkaloid which is composed of the forms A and B (for a review and further references see [132]). The teleocidines differ in chemical structure from the phorbol diesters but are very similar in biochemical and biological activity. Also in this class of chemicals, derivatives like dihydroteleocidin B (the double bond in the lower right corner of the molecule shown below is hydrogenated) are known which are also active tumor promoters. The chemical structures of teleocidin B and A are

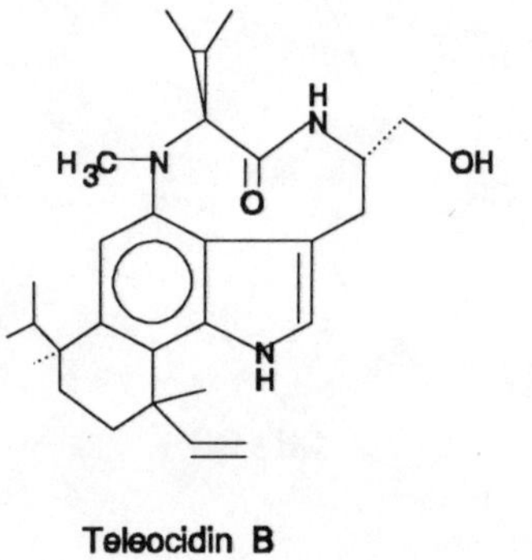

Teleocidin B

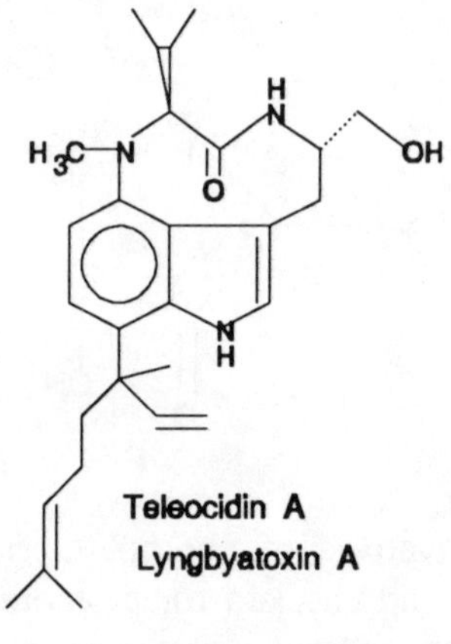

Teleocidin A
Lyngbyatoxin A

Another well known strong tumor promoter is aplysiatoxin isolated from *Lyngbya gracilis*, a kind of seaweed (see again [132]):

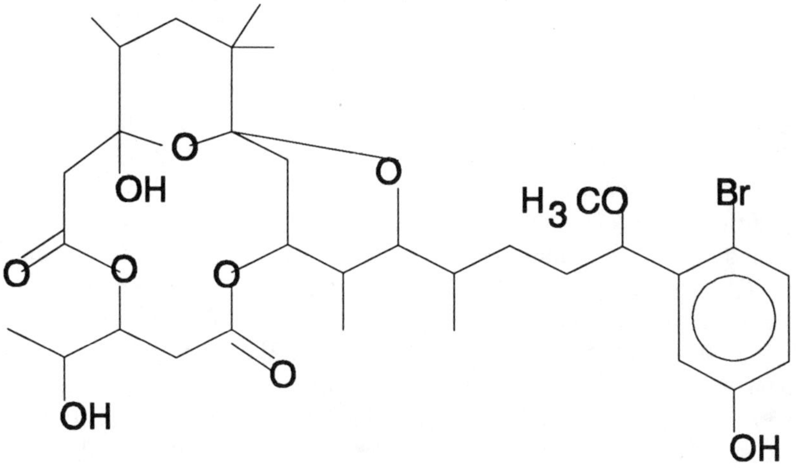

Aplysiatoxin

It is interesting to note that this compound loses most of its promoting activity, but not other biological activities when the bromine is substituted by hydrogen. Some other chemicals which are known to be tumor promoters we just want to list here and refer the reader to [125] for references and details: Fatty acid methyl esters, anthralin, chrysarobin, iodoacetic acid, 7-bromomethylbenz[a]anthracene, benzoyl peroxide, and 2,3,7,8-tetrachlorodibenzo-*p*-dioxin (TCDD), the infamous Seveso poison. Also a fraction of cigarette smoke condensate shows strong promoting activity. Further examples and details can be found in Ref. [124].

In the course of determining the binding site of promoters in the cell membrane, it was first found that all nucleated animal cells which were investigated could bind TPA. Thus the receptor for TPA must be one which is important for the survival of the cell. In 1982 it was shown that TPA can activate a protein kinase from rat brain, called protein kinase C, which depends on the Ca^{2+} and phospholipid concentration. This kinase is widely spread in mammalian tissues. It requires for activity also the presence of diacylglycerol which can be substituted by TPA [133]. Therefore the phorbol diesters bind to protein kinase C sites initially and increase the activity of the enzyme leading to changes in the concentration of phosphoproteins [134,135].

In the remaining part of this section we want to describe briefly the effects which the most prominent and best studied promoters, the phorbol diesters show when they are applied (without initiation by a carcinogen) to cells. For more detailed discussions of these effects and specific references we refer the reader to the literature [124, 125, 136-138]. One of the most obvious effects of phorbol diesters is the reduction of the cell volume by about 20-30%. Also the arrangement of cell cultures in vitro changes in a way similar but not identical to transformed cells. Furthermore it was found that TPA disrupts intercellular communication by an exchange of small molecules, especially it inhibits the metabolic cooperation via cell-cell contact. It was found that normal cells are in extensive communication and become uncoupled by TPA administration, while malignant cells are uncoupled with or without TPA administration. In Ref. [139] a discussion of the role intercellular communication plays in tumor promotion is given. TPA also stimulates the Na^+ and especially the K^+ uptake of cells. This is coupled with a H^+ loss und thus the pH value in the cell is increased or in other words the interior of the cell becomes less acidic ([125], p. 88). TPA stimulates the sugar transport and metabolism in cell cultures as well.

Interestingly TPA also inhibits the binding of the epidermal growth factor (EGF) and of insulin to the cell mebrane. It is known that it does not bind directly to the corresponding receptors, but it is able to delay the appearance of such highly active binding sites in the membrane. This is most probably due to the fact that TPA activates, as already mentioned, the protein kinase C system which can phosphorylate the receptors for these ligands. In this way TPA stimulates DNA synthesis in otherwise dormant cells, e.g. cells with DNA mutations. Also the binding of hormones which release EGF and other important molecules like thyrotropin from the membrane is inhibited by TPA. In this case probably both, the affinity of binding sites and their number is reduced by TPA [140]. Further TPA also stimulates the phospholipid metabolism and the synthesis of prostaglandines in several kinds of cells, including human HeLa cells. In human leukemia cells, TPA also leads to drastic changes in the synthesis of glycoproteins at the cell surface and it was shown that the synthesis of a special glycoprotein which is known to be connected with the malignant transformation is induced by TPA [141]. Also the production of derivatives with anionic oxygen radical groups is enhanced by TPA which is assumed to play a role in tumor promotion [142]. In connection with its role as promoter the effects of TPA stimulated phosphorylation were investigated. It was found that tyrosine phosphorylation in a polypeptide occurs as probably the first step after introduction of TPA into cell cultures. A similar polypeptide is phosphorylated after transformation of cells with Harvey sarcoma virus. This list of examples of reversible changes in cell chemistry by TPA seems to be sufficient for our purpose. However, we want to mention that many other influences of TPA on cells have been reported [125].

Finally we want to comment on the effects TPA has on DNA. First of all TPA stimulates DNA synthesis in several types of cells, including human ones, however, not in all cell strains studied. Consequently, as already discussed the cell population increases, also that of quiescent cells. Further it is reported that TPA can induce sister chromatid exchanges and chromosome aberration, however, it is not clearly known what role this might play in promotion of tumors. A late and maybe irreversible step in tumor promotion might be the observed ability of promoters of several different types to induce gene amplification, which, as we have seen above, plays an important role in the process of oncogene activation. Finally let us mention that Phorbol diesters, by their direct action on the cell surface can also inhibit the cell differentiation of normal and a wide variety of transformed cells [143,144].

Further promoters are also able to affect transformation induced by tumor viruses. The promoters enhance the expression of the transformed phenotype and they also can increase the virus production rate in infected cells or the virus expression in cells which contain a dormant viral genome, thus increasing the rate of transformation (see [145] for a review). It is also known that TPA affects the expression rate of cellular proto-oncogenes, like e.g. c-*myc* or c-*fos*. For more detailed information on tumor promoters and their action in the cell, especially on the phorbol diesters we refer the reader to the review by Diamond [125] and to the work of Weinstein (see for some examples [146-150]). Also for details of the mechanisms of multistage carcinogenesis in general, which is related to tumor promotion the work of Weinstein et al. (see e.g. [151-159]) is recommended.

An interesting class of tumor promoters are the steroids, since they show different activities depending on the dosage in which they are applied. Small amounts of estrogen [160] or progesterone [161,162] given to rats bearing mammary carcinomas e.g. are able to increase the growth of the tumors considerably. On the other hand, if estrogen is administered in large amounts it can inhibit tumor growth. This last effect is augmented by progesterone [163]. Also large doses of androgens can inhibit tumor growth [164,165]. However, for the promoting effect of estrogen to show up, the presence of a functioning pituitary gland is necessary [160]. It was found that the most important pituitary hormone for the promotion is the peptide prolactin [166] which also alone can stimulate tumor growth. However, for a prolonged growth of the tumors, ovarian steroids are necessary [167]. In the case of adrenal steroids much less is known. But it was found that adrenalectomy stimulates the growth of rat mammary carcinomas, however, such a treatment leads also to an increased secretion of prolactin [168,169]. In agreement with these findings the administration of glucocorticoids to rats bearing mammary carcinomas leads to regression of the tumors [170]. However, it is still not clear if the promoting activity is due to glycocorticoid depletion or to the enhanced secretion of prolactin [171]. As just one example for the chemical structures of steroids we show androsterone:

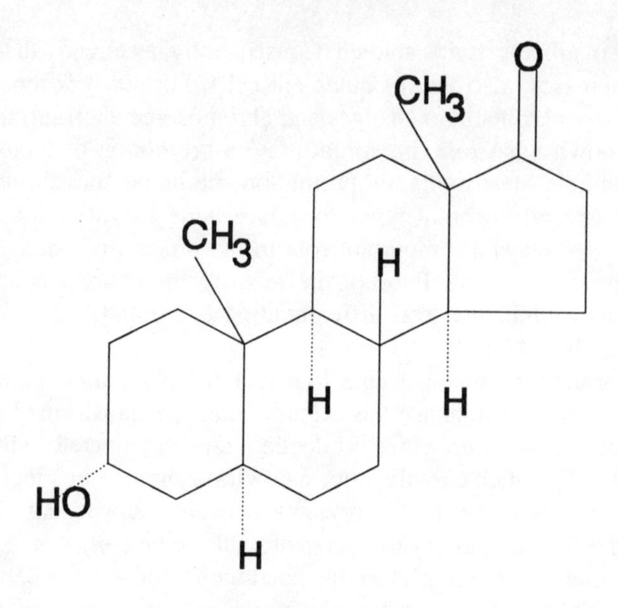

Very recently [172], it has been reported that in case of rat breast cancer high-fat diet has a promoting effect, causing a higher tumor incidence as well as a larger size of tumors compared to a control group with a low-fat diet. As initiating carcinogen dimethylbenzanthracene was applied. In the same study it was also shown that indomethacin has promoting activity, however, in an unusual form. It reduces significantly the number of tumors occurring, which can be viewed as a kind of anti-promoting activity, however, it increases the size of tumors occurring in the high-fat diet group. Mechanisms for these activities are not known yet.

References

1. J. Körbler, "Geschichte der Krebskrankheit" ("The History of Cancer", in German), Verlag Dr. H. Renner, Wien 1973, p. 67.
2. J. Körbler, "Geschichte der Krebskrankheit" ("The History of Cancer", in German), Verlag Dr. H. Renner, Wien 1973, p. 140.
3. K. Yamagiwa and K. Ischikawa, Proc. Med. Fac. Tokyo Univ. **15**, 255 (1916).
4. R. Shapiro and Pohl, Biochemistry **7**, 448 (1968).
5. E. I. Budowsky, Proc. Nucleic Acid Res. Mol. Biol. **16**, 125 (1976).
6. R. F. Kimball, Mutat Res. **39**, 111 (1977).

7. C. Auerbach, M. Moutschen-Dahmen and J. Moutschen, Mutat. Res. **39**, 317 (1977).
8a. P. Daudel and R. Daudel, "Chemical Carcinogenesis and Molecular Biology", John Wiley & Sons, New York, London, Sidney, 1966.
8b. B. Singer and D. Grunberger, "Molecular Biology of Mutagens and Carcinogens", Plenum Press, New York and London 1983.
9. M. Barbacid, Ann. Rev. Biochem. **56**, 779 (1987).
10. K. F. Muench, R. P. Misra and M. Z. Humayun, Proc. Natl. Acad. Sci. USA **80**, 6 (1983).
11. J. A. Hartley, N. W. Gibson, K. W. Kohn, and W. B. Mathes, Cancer. Res. **46**,1943 (1986).
12. D. Toorchen and M. D. Topal, Carcinogenesis **4**, 1591 (1983).
13. M. D. Topal, J. S. Eadie and M. Conrad, J. Biol. Chem. **261**, 9879 (1986).
14. H. Druckrey, S. Ivankovic and J. Gimmy, Z. Krebsforsch. **79**, 282 (1973).
15. D. B. Ludlum and W. P. Tong, in "Nitrosoureas: Current Status and New Developments", eds. A. W. Prestayko, S. T. Crooke, L. H. Baker, S. K. Carter, and P. S. Schein, Academic Press, New York 1981, pp. 85.
16. R. Montesano and H. Bartsch, Mutat. Res. **32**, 179 (1976).
17. A. E. Pegg, Adv. Cancer Res. **5**, 195 (1977).
18. D. Kim and F. P. Guenreich, Carcinogenesis **11**, 419 (1990).
19. P. L. Grover (ed.), "Chemical Carcinogens and DNA", Vol. I, CRC Press, Inc., Boca Raton, Florida, 1979.
20. P. L. Grover (ed.), "Chemical Carcinogens and DNA", Vol. II, CRC Press, Inc., Boca Raton, Florida, 1979.
21. W. Hueper and W. D. Conway, in "Chemical Carcinogenesis and Cancer", C. C. Thomas (ed.), Springfield, Illinois, 1964.
22. R. Freudenthal and P. W. Jones, "Chemical Carcinogenesis - A Comprehensive Study", Vol. 1, Ragen, New York, 1976.
23. W. C. Herndon, Intern. J. Quantum Chem. **QBS1**, 123 (1974).
24. I. B. Weinstein, A. M. Jeffrey, K. W. Jennbette, S. H. Blobstein, R. G. Harvey, C, Harris, A. Autrup, K. Kasa, and K. Makanish, Science **197**, 92 (1977).
25. K. Nakanishi, H. Kasai, H. Cho, R. G. Harvey, A. M. Jeffrey, K. W. Jennette, and I. B. Weinstein, J. Amer. Chem. Soc. **99**, 258 (1977).
26. H. Yagi, H. Akagi, D. R. Thakker, H. D. Mah, M. Koreeda, and D. M. Jerina, J. Amer. Chem. Soc. **99**, 2358 (1977).
27. R. Lavery, A. Pullman and B. Pullman, Intern. J. Quantum Chem. **QBS5**, 21 (1978).
28. D. M. Jerina and J. W. Daly, in "Drug Metabolism", D. V. Parke and R. L. Smith (eds.), Taylor and Francis, London, 1976.
29. B. Pullman, in "Hydrocarbons and Cancer", H. V. Gelboin (ed.), Academic Press, New York, 1973.

30. A. Pullman and B. Pullman, Adv. Cancer Res. **3**, 117 (1953).
31. Levin et al., J. Biol. Chem. **255**, 9067 (1980).
32. H. V. Gelboin, Physiol. Rev. **60**, 1107 (1980).
33. J. DiGiovanni, J. F. Sina, S. W. Ashurst, J. M. Singer, and L. Diamond, Cancer Res. **43**, 163 (1983).
34. P. Sims and P. L. Grover, in "Polycyclic Hydrocarbons and Cancer", H. V. Gelboin and P. O. P. Ts'o (eds.), Academic Press, New York, pp. 117 (1981).
35a. I. B. Weinstein, A. M. Jeffrey, K. W. Jennette, S. H. Blobstein, R. G. Harvey, C. Harris, H. Autrup, K. Kasa, and K. Makanish, Science **197**, 92 (1977); K. Nakanishi, H. Kasai, H. Cho, R. G. Harvey, A. M. Jeffrey, K. W. Jennette, and I. B. Weinstein, J. Amer. Chem. Soc. **99**, 258 (1977); H. Yamasaki, P. Pulbrabeck, D. Grunberger, and I. B. Weinstein, Cancer Res. **37**, 3756 (1977).
35b. P. Otto, J. Ladik and S. C. Liu, J. Mol. Struct. (Theochem) **123**, 129 (1985).
36. T. F. Slater and P. A. Riley, Intern. J. Quantum Chem. **QBS5**, 173 (1978).
37. J. Ladik, P. Otto and W. Förner, Intern. J. Quantum Chem. **QBS10**, 73 (1984).
38. W. Förner, P. Otto and J. Ladik, Chem. Phys. **86**, 49 (1984).
39. P. Otto, Intern. J. Quantum Chem. **30**, 275 (1986).
40. E. Kriek, J. A. Miller, U. Juhl, and E. C. Miller, Biochem. **6**, 177 (1967).
41. H. Yamasaki, P. Pulkrabeck, D. Grunberger, and I. B. Weinstein, Cancer Res. **37**, 3756 (1977).
42. G. P. Margison and P. J. O'Connor, Biochim. Biophys. Acta Libr. **331**, 349 (1973).
43. E. Kriek and J. G. Westra, in "Chemical Carcinogens and DNA", Vol. II, P. L. Grover (ed.), CRC Press Inc., Boca Raton, Florida, pp. 1 (1979).
44. S. S. Thorgairsson, P. J. Wirth, N. Staiano, and C. L. Smith, Adv. Exp. Med. Biol. **136B**, 897 (1982).
45. G. N. Wogan, J. M. Essigmann, R. G. Croy, W. F. Busby, J. D. Groopman, and A. A. Stark, in "Naturally-Occurring Carcinogens: Mutagens and Modulators of Carcinogenesis", E. C. Miller, J. A. Miller, I. Hirono, T. Sugimura, and S. Takayama (eds.), University Park Press, Baltimore, pp. 19 (1979).
46. E. Boyland, in "Proceedings of the Israel Academy of Sciences: Symposium on Carcinogenesis", Jerusalem, 1968; H. E. Kaiser, in "Cancer Growth and Progression", Vol. 2: "Mechanisms of Carcinogenesis", Ed. E. K. Weisburger, Kluwer Academic Publishers (Dordrecht, Boston, London, 1989), pp. 19.

47. R. Knippers, P. Phillipsen, K. P. Schäfer, and E. Fanning, "Molekulare Genetik" (in German), Georg Thieme Verlag, Stuttgart - New York, 1990, p. 299.
48. D. A. Youngs and K. C. Smith, Mutat Res. **51**, 133 (1978).
49. A. Sancar and C. S. Rupert, Mutat. Res. **51**, 139 (1978)
50. R. M. Snapka and B. M. Sutherland, Biochemistry **19**, 4201 (1980).
51. C. Helene, M. Charier, J. Toulme, and F. Toulme, in "DNA Repair Mechanisms", E. C. Friedberg and C. F. Fox (eds.), Academic Press, New York, 1978, pp. 141.
52. B. M. Sutherland, see Ref. 51., pp. 113.
53. T. Lindahl, Ann. Rev. Biochem. **51**, 61 (1982).
54. E. Seeberg, Proc. Natl. Acad. Sci. USA **75**, 2569 (1978).
55. E. Seeberg, Mutat. Res. **82**, 11 (1981).
56. E. Seeberg, Proc. Nucl. Acid Res. Mol. Biol. **26**, 217 (1981).
57. G. H. Yoakum, A. Sancar and W. D. Rupp, Nucleic Acids Res. **9**, 4495 (1981).
58. E. M. Witkin, Bacteriological Rev. **40**, 869 (1976).
59. G. C. Walker, Microbiol. Rev. **48**, 60 (1982).
60. J. J. Weigle, Proc. Natl. Acad. Sci. USA **39**, 628 (1953).
61. E. M. Witkin, Science **152**, 1345 (1966).
62. E. M. Witkin, Ann. Rev. Microbiol. **23**, 487 (1969).
63. B. A. Bridges, Ann. Rev. Nucl. Sci. **19**, 139 (1969).
64. S. Kondo, Genetics **73**, 109 (1973).
65. C. Auerbach, "Mutation Research", Halsted Press, New York, 1976.
66. C. O. Doudney, in "Photochemistry and Photobiology of Nucleic Acids", Vol. II, S. Y. Wang (ed.), Academic Press, New York (1976), pp. 309.
67. J. W. Drake, "The Molecular Basis of Mutation", Holden-Day, San Francisco, 1970.
68. J. W. Drake and R. H. Baltz, Ann. Rev. Biochem. **45**, 11 (1976).
69. C. W. Lawrence and R. Christensen, Genetics **82**, 207 (1976).
70. D. E. Brash and W. A. Haseltine, Nature **298**, 189 (1982).
71. P. W. Doetsch, L. G. Chan and W. A. Haseltine, Nucleic Acids Res, **13**, 3285 (1985).
72. S. N. Rose, R. J. H. Davies, S. K. Sethi, and J. A. McClosky, Science **220**, 723 (1983).
73. S. Y. Wang, "Photochemistry and Photobiology", Vol. 2, Academic Press, New York, 1976.
74. P. S. Song and K. J. Tapley, Photochem. Photobiol. **29**, 1177 (1979).
75. D. Kanne, K. M. Straub, J. E. Hearst, and H. Rapoport, J. Amer. Chem. Soc. **104**, 6754 (1982).
76. D. Kanne, K. M. Straub, H. Rapoport, and J. E. Hearst, Biochemistry **21**, 861 (1982).

77. A. C. Upton, Sci. Am. 246, 29 (1982).
78. T. Vogt and J. Lissner, Deutsches Ärzteblatt 82, 1169 (1986).
79. S. H. Revell, Adv. Radiat. Biol. 4, 367 (1974).
80. J. F. Ward, Prog. Nucleic Acid Res. Mol. Biol. 35, 95 (1988).
81. W. R. Russel and E. M. Kelly, Proc. Natl. Acad. Sci. USA 79, 542 (1982).
82. C. Waldren, L. Correl, M. A. Sognier, and T. T. Puck, Proc. Natl. Acad. Sci. USA 83, 4839 (1986).
83. K. Sax, Genetics 23, 494 (1938).
84. K. Sax, Cold Spring Harbor Symp. Quant. Biol. 9, 93 (1941).
85. R. J. Preston, Environm. Mol. Mutagenesis 14, 126 (1989).
86. D. E. Lea, "Actions of Radiations on Living Cells", Cambridge Univ. Press, London and New York, 1955.
87. D. E. Lea and D. G. Catcheside, J. Genet. 44, 216 (1942).
88. J. M. Thoday, J. Gent. 43, 189 (1942).
89. J. M. Thoday, Brit. J. Radiol. 24, 572, 622 (1951); A. C. Upton, Cancer Res. 21, 717 (1961); A. C. Upton, Radiation Res. 71, 51 (1977); A. C. Upton, in "Cancer Growth and Progression", Vol. 2: "Mechanisms of Carcinogenesis", Ed. E. K. Weisburger, Kluwer Academic Publishers (Dordrecht, Boston, London, 1989), pp. 54;
90. L. J. Stadler, Proc. 6th Int. Congr. Genet. 1, 277 (1932).
91. S. H. Revell, in "Radiobiology Symposium, Liege", Z. M. Bacq and P. Alexander (eds.), Butterworth (London 1955), pp. 243.
92. S. H. Revell, Ann. N. Y. Acad. Sci. 68, 802 (1958).
93. S. H. Revell, Proc. Roy. Soc. London Ser. B 150, 563 (1959).
94. A. M. V. Duncan and H. J. Evans, Mutat. Res. 107, 307 (1983).
95. J. R. K. Savage, R. J. Preston and G. J. Neary, Mutat. Res. 5, 47 (1968).
96. A. C. Van Kesteren-van Leeuwen and A. T. Natarajan, Chromosoma 81, 473 (1980).
97. A. T. Natarajan and G. Obe, in "Radiation-Induced Chromosome Damages in Man", T. Ishihara and M. S. Sasaki (eds.), Alan R. Liss (New York, 1983), pp. 127.
98. A. T. Natarajan and G. Obe, Chromosoma 90, 120 (1984).
99. R. J. Preston, loc. cit. 97., pp. 111.
100. R. E. Merrifield, J. Chem. Phys. 28, 647 (1958).
101. N. Wiser, Phys. Rev. 129, 62 (1963).
102. J. Ladik, J. Mol. Struct. (Theochem) 277, 109 (1992).
103. W. F. Malone, in "Mechanisms of Carcinogenesis", E. K. Weisburger (ed.), Kluwer Academic Publishers, Dordrecht, Boston, London, 1989, p. 32.
104. H. E. Kaiser, loc. cit. 103., pp. 1.

105. H. E. Kaiser, loc. cit. 103., pp. 19.
106. A. C. Upton, loc. cit. 103., p. 63.
107. S. Milham, N. Engl. J. Med. **307**, 249 (1982).
108. N. Wertheimer and E. Leeper, Intern. J. Epidemiol. **11**, 345 (1982).
109. N. Wertheimer and E. Leeper, Ann. N. Y. Acad. Sci. **502**, 43 (1987).
110. W. F. Friedewald and P.Rous, J. Exp. Med. **8**, 101 (1944).
111. J. C. Mottram, J. Pathol. Bacteriol. **56**, 181 (1944).
112. I. Berenblum and P. Shubik, Br. J. Cancer **1**, 379 (1947); ibid. **3**, 384 (1949).
113. I. Berenblum, Adv. Cancer Res. **2**, 129 (1954).
114. B. L. VanDuuren and L. Orris, L. Cancer Res. **25**, 1871 (1965).
115. B. L. VanDuuren, Prog. Exp. Tumor Res. **11**, 31 (1969).
116. R. K. Boutwell, Prog. Exp. Tumor Res. **4**, 207 (1964).
117. T. J. Slaga, S. M. Fischer, C. E. Weeks, K. Nelson, M. Mamrack, and A. J. P. Klein-Szanto, in "Carcinogenesis - A Comprehensive Survey", Vol. 7, E. Hecker, W. Kunz, N. E. Fusenig, F. Marks, and H. W. Thielmann (eds.), Raven Press, New York, 1982, pp.19.
118. T. J. Slaga, in "Modulation and Mediation of Cancer by Vitamins", F. L. Maiskens and K. N. Prasad (eds.), S. Karger, Basel, 1983, pp. 10.
119. P. Rous and J. G. Kidd, J. Exp. Med. **73**, 365 (1941).
120. V. Potter, Yale J. Biol. Med. **53**, 367 (1980).
121. V. Potter, Carcinogenesis **2**, 1375 (1981).
122. C. Peraino, R. J. M. Fry and E. Staffeldt, Cancer Res. **31**, 1506 (1971).
123. C. Peraino, R. J. M. Fry, E. Staffeldt, and W. E. Kisieleski, Cancer Res. **33**, 2701 (1973).
124. L. Diamond, T. G. O'Brien and and W. M. Baird, Adv. Cancer Res. **42**, 1 (1980).
125. L. Diamond, in "Mechanisms of Cellular Transformation by Carcinogenic Agents", D. Grunberger and S. Goff (eds.), Pergamon Press, Oxford, New York, 1987, pp. 73.
126. E. Hecker, W. Kunz, N. E. Fusenig, F. Marks, and H. W. Thielmann (eds.), "Carcinogenesis - A Comprehensive Survey" Vol. 7, Raven Press, New York, 1982.
127. T. J. Slaga, "Mechanisms of Tumor Promotion", Vol. 1-4, CRC Press, Inc., Boca Raton, Florida, 1983, 1984.
128. E. Hecker, in "Carcinogenesis", Vol.2, T. J. Slaga, A. Sivak and R. K. Boutwell (eds.), Raven Press, New York, 1978, pp.11.
129. E. Hecker, in "Cancer Research", Vol. 6, H. Bush (ed.), Academic Press, New York, 1971, pp. 439.
130. T. J. Slaga, S. M. Fischer, K. Nelson, and G. L. Gleason, Proc. Natl. Acad. Sci. USA **77**, 3659 (1980).

131. G. Fürstenberger, H. Richter, T. S. Argyris, and F. Marks, Cancer. Res. **42**, 342 (1982).
132. G. Sugimura, Gann **73**, 499 (1982).
133. M. Castagna, Y. Takai, K. Kaibuchi, K. Sano, U. Kikkawa, and Y. Nishizuka, J. Biol. Chem. **257**, 7847 (1982).
134. I. B. Weinstein, Nature **302**, 750 (1983).
135. Y. Nishizuka, Nature **308**, 693 (1984).
136. P. M. Blumberg, CRC Crit. Rev. Tox. **8**, 153, 199 (1980).
137. A. M. Mastro, Lymphokines **6**, 263 (1982).
138. A. M. Mastro, Cell. Biol. Int. Rep. **7**, 881 (1983).
139. J. E. Trosko, C.-C. Chang and A. Metcalf, Cancer Invest. **1**, 511 (1983).
140. B. D.-M. Chen, H. S. Lin and S. Hsu, J. Cell. Physiol. **116**, 207 (1983).
141. M. M. Gottesmann and S. H. Yuspa, Carcinogenesis **2**, 971 (1981).
142. C. S. Baxter, in "Mechanisms of Tumor Promotion", T. J. Slaga (ed.), CRC Press, Boca Raton, Florida, 1984.
143. R. Cohen, M. Pacifici, N. Rubinstein, J. Biehl and H. Holzer, Nature **266**, 538 (1977).
144. S. H. Yuspa, in "Normal and Abnormal Epidermal Differentiation", M. Seiji and I. A. Bernstein (eds.), Proc. Japan-US Seminaron Normal and Abnormal Epidermal Differentiation, University of Tokyo Press, Tokyo, 1983, pp. 227.
145. N. Yamamoto, Rev. Physiol. Biochem. Pharmacol. **101**, 111 (1984).
146. I. B. Weinstein, M. Wigler and C. Pietropaolo, in "The Origins of Human Cancer", Eds. H. H. Hiatt, J. D. Watson and J. A. Winstein, Cold Spring Harbour Conferences on Cell Proliferation, Vol. 4, Cold Spring Harbour, New York, 1977, pp. 751.
147. I. B. Weinstein, R. A. Mufson, L. S. Lee, P. B. Fisher, J. Laskin, A. D. Horovitz, and V. Ivanovic, in "Carcinogenesis: Fundamental Mechanisms and Environmental Effects", Eds. P. Pullman, P. O. P. Ts'o and H. Gelboin, D. Reidel Publ. Comp., Dordrecht, Holland, 1980, pp. 543.
148. I. B. Weinstein, Nature, **302**, 750 (1983).
149. I. B. Weinstein, in "Carcinogensis", Vol. 10, Eds. E. Huberman and S. H. Barr, Raven Press, New York, 1985, pp. 177.
150. W. L. W. Hsiao, S. Gattoni-Celli and I. B. Weinstein, Science **226**, 552 (1984).
151. I. B. Weinstein, S. Gattoni-Celli, P. Kirschmeier, M. Lambert, W. L. W. Hsiao, J. Backer, and A. Jeffrey, Cancer Cells **I**, 229 (1984).
152. I. B. Weinstein, J. Arcoleo, M. Lambert, W. L. W. Hsiao, S. Gattoni-Celli, A. M. Jeffrey, and P. Kirschmeier, in "Molecular Biology of Tumor Cells", Ed. B. Wahren, Raven Press, New York, 1985, pp. 55.

153. S. Gattoni-Celli, W. L. W. Hsiao, M. Lambert, P. Kirschmeier, and I. B. Weinstein, in "Carcinogenesis", Vol. 9, Eds. R. W. Tennant and J. C. Barrett, Raven Press, New York, 1985.
154. I. B. Weinstein, J. Cell. Biochem. **33**, 213 (1987).
155. I. B. Weinstein, Cancer Res. **48**, 4135 (1988).
156. I. B. Weinstein, Mutat. Res. **202**, 413 (1988).
157. I. B. Weinstein, Cancer Detect. Prevent. **14**, 253 (1989).
158. I. B. Weinstein, Cancer Res. (Suppl.) **51**, 5080s (1991).
159. I. B. Weinstein, Science **251**, 387 (1991).
160. A. Sterental, J. M. Dominguez, C. Weissman, and O. H. Pearson, Cancer Res. **23**, 481 (1963).
161. A. G. Jabara, Brit. J. Cancer, **21**, 418 (1967).
162. G. M. McCormick and R. C. Moon, Cancer Res. **27**, 626 (1967).
163. C. Huggins, R. C. Moon and S. Morii, Proc. Natl. Acad. Sci. USA **48**, 379 (1962).
164. A. Danguy, N. Legros, N. Devleeschouwer, J. A. Heuson-Stennon, and J. C. Heuson, in "Role of Methoxyprogesterone in Endocrine Related Tumors", Eds. S. Iacobelli and A. DiMarco, Reaven Press, New York, pp. 21.
165. S. Young, R. A. Baker and J. E. Helfenstein, Brit. J. Cancer **19**, 155 (1965).
166. C. W. Welsch and H. Nagasawa, Cancer Res. **37**, 951 (1977).
167. D. Sinha, D. Cooper and D. L. Dao, Cancer Res. **33**, 411 (1973).
168. C. F. Aylsworth, C. A. Hodson, G. Berg, G. Kledzik, and J. Meites, Cancer Res. **39**, 2436 (1979).
169. H. J. Chen, C. J. Bradley and J. Meites, Cancer Res. **36**, 1414 (1976).
170. C. F. Aylsworth, P. W. Sylvester, F. C. Leung, and J. Meites, Cancer Res. **40**, 1863 (1980).
171. C. W. Welsch, in "Mechanisms of Carcinogenesis", E. K. Weisburger (ed.), Kluwer Academic Publishers, Dordrecht, Boston, London, 1989, pp. 169.
172. Y. Mizukami, M. Noguchi, A. Nonomura, T. Taniya, and N. Ohta, Anticancer Res. **12**, 1847 (1992).

4 A Brief Summary of the Structure of DNA and Proteins and their Biological Role

4.1 The Chemical and Geometrical Structure of DNA and Proteins

Avery, McLeod and McCarty [1] discovered in 1944 that the giant molecule DNA, which had been known since the second half of the last century (and which occurs primarily in the nuclei of the cells), is the carrier of the genetic message. Since then DNA has been the subject of very intensive studies. Its chemical structure has been determined, that is what kind of molecules build it up and in which way these molecules are linked to each other. Further with the help of X-ray diffraction they also established its geometrical structure, in other words what is the overall shape of this macromolecule and where are precisely in space those atomic nuclei which build up the constituent molecules of DNA.

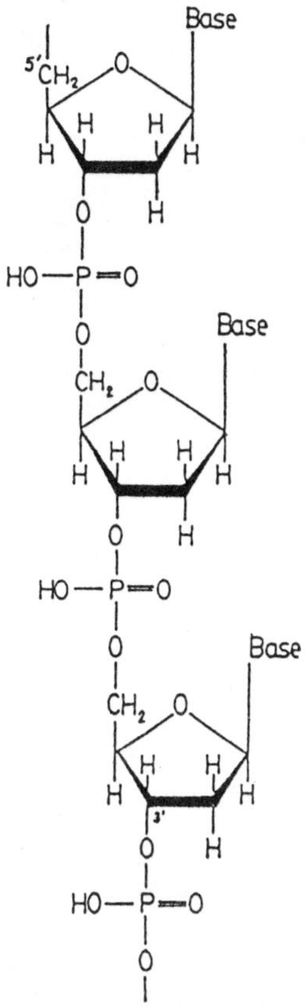

Fig. 4.1: The overall chemical arrangement of a DNA chain

The chemical work showed that DNA is a long chain of molecules consisting of a sugar-phosphate backbone to which the four nucleotide bases are attached [2]. Fig. 4.1 presents the overall chemical arrangement of a DNA chain, while Fig. 4.2 shows the chemical formulae of these four bases: adenine (A), thymine (T), guanine (G) and cytosine (C).

CYTOSINE (C)

THYMINE (T)

ADENINE (A)

GUANINE (G)

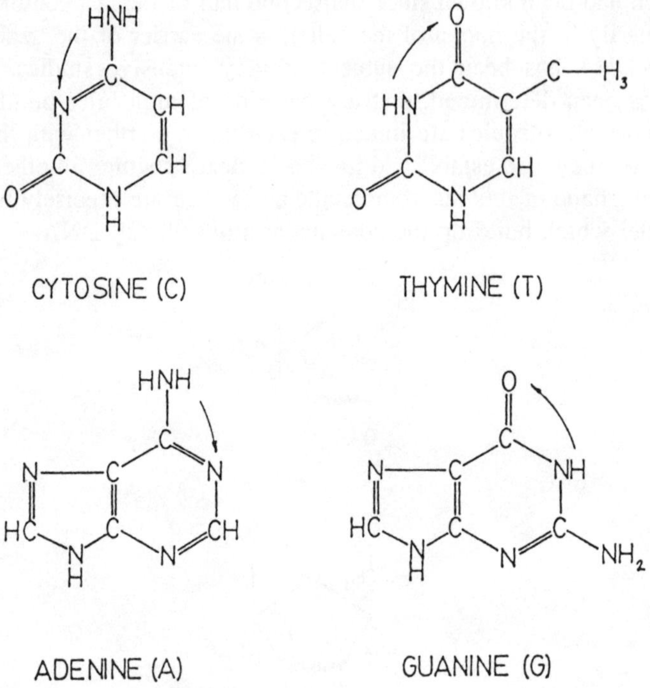

Fig. 4.2: The chemical formulae of the four nucleotide bases: adenine (A), thymine (T), guanine (G) and cytosine (C). The *arrows* indicate those hydrogen atoms which can be shifted to another position

While the sugar-phosphate backbone is repetitive in its units, the four bases follow each other generally in a non-periodic way in the chain. The DNA of a virus is only a few ten thousand units long (the number of nucleotide bases in its chain), while the DNA of a higher organism is much longer. For example the DNA content of a human cell consists of about 10^{10} bases. If we consider the four different molecules occurring in DNA, A, T, G, C, as letters of a four-letter alphabet, we can say that the genetic message is coded in DNA through the sequence of these four letters. For instance the 9 letter sequence AATGATCAG has a completely different genetic meaning another 9 letter sequence GACAAGATT (though in these examples even the number of the different molecules is the same, i.e. we have 4 A-s, 2 G-s, 2 T-s and 1 C). The reader should remember that even the largest electronic computers which can perform extremely complicated calculations, use only the two-letter alphabet of 0 and 1 (binary code). We shall see later what the probable reason is for nature choosing a four-letter alphabet in the course of biological evolution instead of the simplest two-letter one.

Fig. 4.3 shows the detailed chemical structure of a sugar-phosphate group of DNA which forms the link between the different units of this chain. The two small arrows show the positions where this group is bound to the neighbouring sugar-phosphate unit and the large arrow indicates the carbon atom which is bound to the N-atoms of the bases.

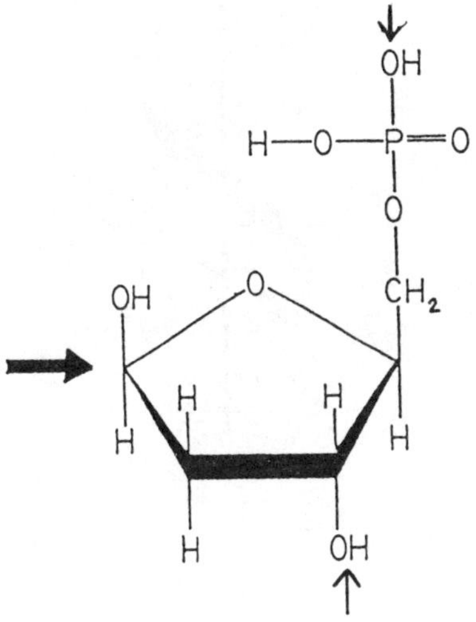

Fig. 4.3: The sugar-phosphate unit of DNA. In the five-membered ring of sugar there is one O-atom and four C-atoms (the latter are not indicated explicitly in the Figure)

On the basis of the detailed X-ray work on the fibers of the potassium salt of DNA done by Wilkins and his coworkers [3] Watson and Crick established in

1953 a geometrical model for DNA [4]. This model (called nowadays DNA B which is the most frequent form of DNA in the living cell) has been proven to be basically correct even when confronted with the results of subsequent, more refined X-ray diffraction investigations of DNA [5].

According to the Wilkins-Watson-Crick model of DNA, this macromolecule forms a double helix in which the sugar-phosphate backbones are outside and the nucleotide bases (which are nearly perpendicular to the main axis of the double helix) are inside (see Fig. 4.4).

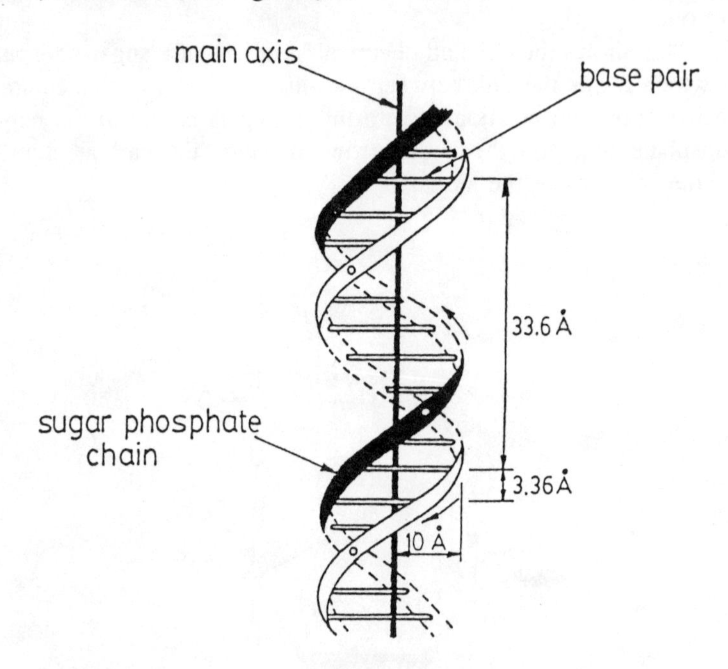

Fig. 4.4: The Wilkins-Watson-Crick double helix model of DNA B (schematic)

The overall diameter of this double helix is about 20 Å. Both helices in the double helix are right-handed and they run in opposite directions. Since A and G are bulkier than T and C, to accommodate two bases in the same plane perpendicular to the main axis of the double helix and to keep at the same time the overall diameter 20 Å, one must always combine either of A or G with either of T and C. In reality, as experimentally shown by X-ray analysis and theoretical considerations, A is always paired to T and G to C. These base pairing relations A-T and G-C are very important in the duplication of DNA (see below). In Fig.

4.5 these so-called Watson-Crick base pairs are presented. The dotted lines represent hydrogen bonds.

The hydrogen bonds are weak bonds (their strength is usually less than one tenth of a normal chemical bond) and occur if a hydrogen atom, H, is situated between two atoms A and B which strongly attract electrons. This H-atom is usually strongly bound to one of the atoms which takes away a part of its electric charge and therefore it will be attracted also by the other "electron hungry" atom. We can write down this situation symbolically as $A^{\delta-}$-$H^{\delta+}$...$B^{\delta-}$, where the δ+ and δ- signs indicate a fraction of an elementary positive or negative charge, respectively.

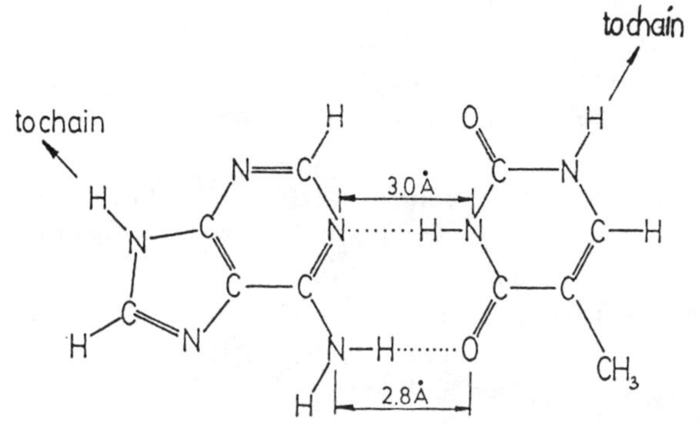

ADENINE-THYMINE BASE PAIR

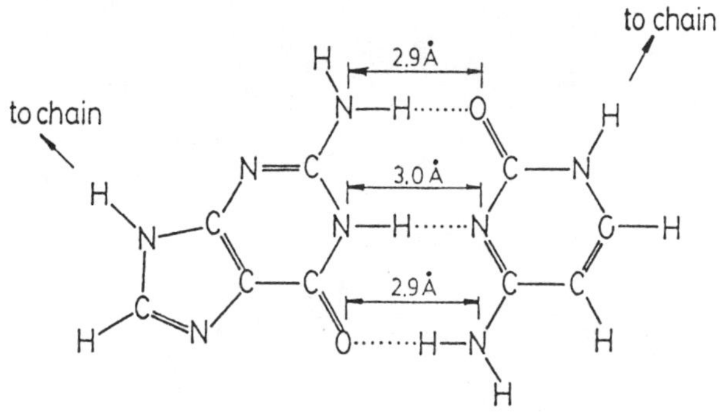

GUANINE-CYTOSINE BASE PAIR

Fig. 4.5: The A-T and G-C Watson-Crick base pairs (schematic). The *dotted lines* represent hydrogen bonds. The *arrows* indicate those positions where the nucleotide bases are bound to the sugar molecules

In the nucleotide bases there is a good possibility for the formation of such hydrogen bonds between an oxygen (O) and a nitrogen (N) atom (both atoms attract electrons) or between two N-atoms. We can easily understand from Fig. 4.5 that A can easily form two hydrogen bonds with T, but not with C, because in the latter case two H-atoms would occur in the same bond which would lead to a strong repulsion. Similarly G can form three hydrogen bonds with C but for the same reason as in the case of A, none with T.

The A-T, G-C base pairing relations mean that if we know the sequence of the nucleotide bases in one helix of DNA, on its basis we can immediately write done the sequence in the complementary other chain. At the same time due to these relations we obtain the ratios [A]/[T] = [G]/[C] = 1 where the square brackets indicate the corresponding quantities, for instance [C] is the quantity of C in a given DNA macromolecule. On the other hand the ratios [A]/[G] = [T]/[C] vary (depending on the biological source of the investigated DNA) between 0.5 and 2.5.

To complete the description of the geometrical structure of DNA according to the Wilkins-Watson-Crick model, the double helix has a full turn after every 33.6 Å and there are ten base pairs in each turn (see Fig. 4.4). This means that the distance between the stacked bases (the so-called stacking distance) is 3.36 Å and one base is rotated around the helix axis by 36° in an anticlockwise way as compared to the previous one, if we look at the double helix from below (see Fig. 4.6).

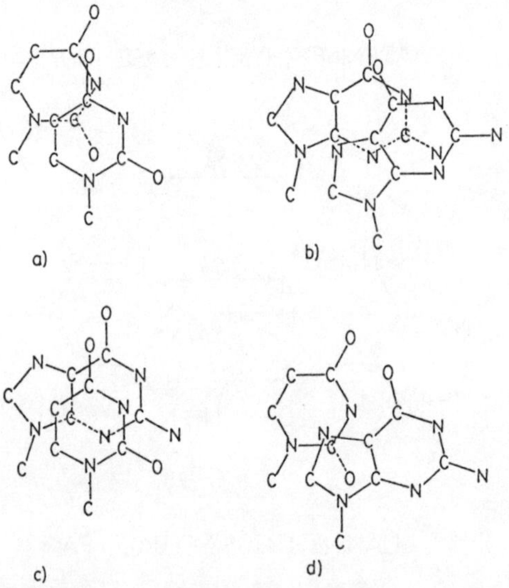

a) b)

c) d)

Fig. 4.6: The relative positions of the stacked bases in DNA B: (a) stacked TT, b) GG, c) GT, d) TG (note that B'B always indicates stacked bases B' and B with B' below B)

It is interesting that this stacking distance in DNA is the same as the distance between adjacent layers in graphite. Note that by definition we indicate by B'-B always hydrogen bonded bases B' and B, while B'B means stacked bases B' and B, with B' below B.

Even in the early investigations it was found that, if the relative humidity of the environment is below 70%, DNA goes from its in vivo (in the cell) stable geometrical arrangement (which was just briefly described above) over to another one, in which the base pairs are more tilted with respect to a perpendicular position to the main axis of the double helix than in DNA B. Since this more tilted configuration of the bases in DNA (which is called DNA A) is not important from the biological point of view, we will not describe it in more detail.

Subsequently much more refined X-ray studies of DNA at different research centers have revealed a whole family of conformations of this molecule. Since most of them do not seem to have larger biological significance, we will not discuss them here, with one exception and this is Z DNA. Z DNA which also occurs inside the cell was discovered by Rich and his coworkers [6] in 1980. It is stable against DNA B only if there is a sequence

$$\begin{matrix} G & C & G & C \\ | & | & | & | \\ C & G & C & G \end{matrix}$$

present in the DNA double helix with a length of at least 4 base pairs. Even such sequences can remain in a DNA B configuration, depending on the concentrations of ions around the DNA. A change of the ionic distribution around DNA or interactions with certain carcinogens like AAF (see Chapter 3), however, can push it over from the B to the Z form.

Z DNA has, in contrast to DNA B not one, but two base pairs as units:

$$\begin{pmatrix} G & - & C \\ C & - & G \end{pmatrix}$$

Furthermore, in the double helix, both helices are left-handed. Though the overall chemical arrangement (sugar-phosphate backbone with the nucleotide bases bound to the sugar and Watson-Crick base pairing) remains very similar to DNA B, the relative positions of the DNA constituents are rather different (see Fig. 4.7). Further instead of ten, Z DNA has twelve base pairs per turn and the double helix is somewhat slimmer (the diameter is 18 Å instead of 20 Å as in DNA B).

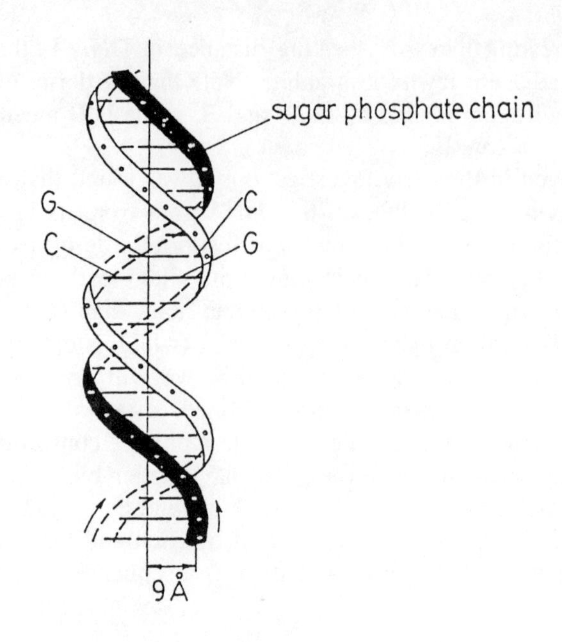

<u>Fig. 4.7</u>: Overall view of Z DNA (schematic)

Due to this changed geometrical structure the nucleotide base pairs are much less shielded against the attack of chemicals or possible effects of radiation. For this reason, many scientists believe that DNA segments in the Z conformation are the long-postulated so-called "hot spots" where damage to DNA occurs more easily.

Nowadays it is well established experimentally [7] that DNA duplicates in a way that the double helix starts to unwind at a certain point, usually due to the action of a certain protein catalyst (enzyme). A catalyst is a larger molecule in solution or a solid on whose surface a chemical reaction takes place. In the process of catalysis (the action of the catalyst) the reaction partners are first bound to the surface which causes such a rearrangement of their electrons that the reaction takes place much more easily (with higher probability). After the completion of the reaction, the products leave the surface of the catalyst which is then ready for action again. To the unpaired single helices formed in this way new helices are built in a way that the A-T and G-C base pairing relations are observed (again with the help of certain enzymes). In this way finally instead of the original double helix two new double helices are formed, each one containing one chain of the original double helix and a new complementary one (semi-conservative duplication). For instance if in the original double helix we had the sequence

```
 A G C C T T A C G
-| | | | | | | | |-
 T C G G A A T G C
```

we obtain from this after duplication the two new double helices:

```
   T C G G A A T G C     new helix
-  | | | | | | | | |  -
   A G C C T T A C G     old helix
```

and

```
   T C G G A A T G C     old helix
-  | | | | | | | | |  -
   A G C C T T A C G     new helix
```

Fig. 4.8 shows in a schematic way the described duplication of DNA.

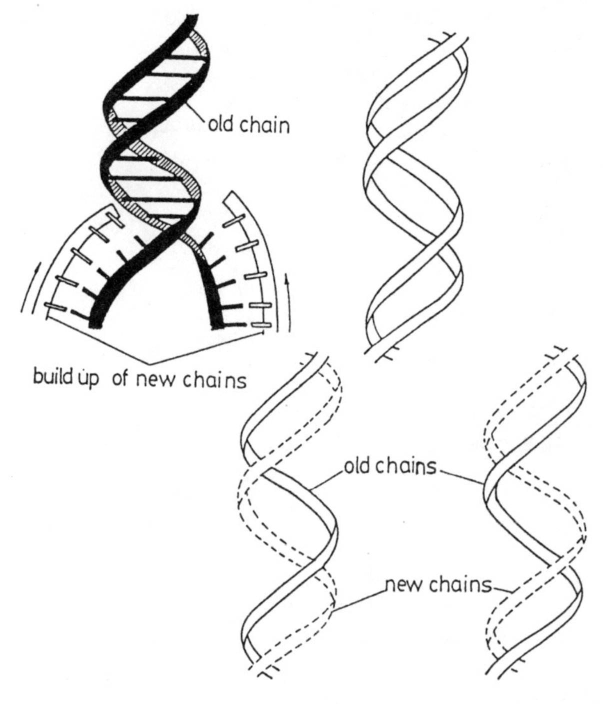

Fig. 4.8: The duplication of the DNA double helix (schematic)

If one looks at Fig. 4.2, it is easy to see that if one moves the H-atom indicated by an arrow, the resulting new forms of A, T, G and C (which we shall denote in the future by A^*, T^*, G^*, C^*) cannot form hydrogen bonds with their original partners. As Watson and Crick [4] have pointed out in these cases the new base pairing relations A^*-C, G^*-T, A-C^*, G-T^* (see Fig. 4.9) are valid.

a.)

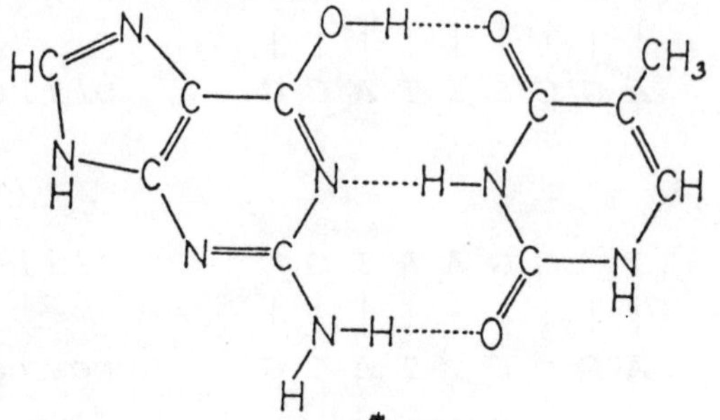

G^*-T

b.)

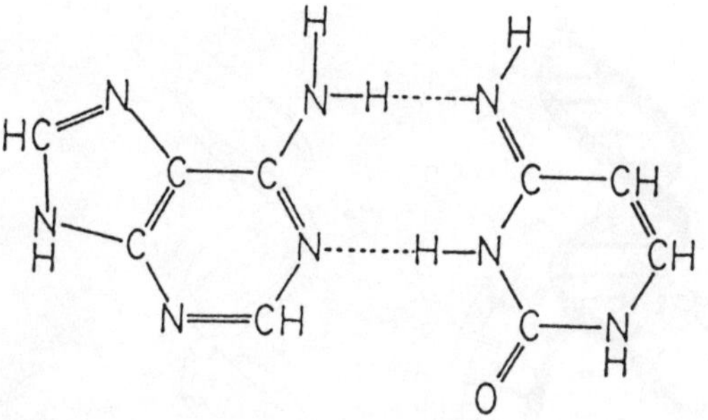

A-C^*

Fig. 4.9: New base pairs in the case of a shift of a H-atom in one of the nucleotide bases.
a) G^*-T, b) A^*-C

If this shift of the H-atoms occurs during DNA duplication, the newly formed chain will have a permanent change in its sequence. For instance if in the above given example after the opening of the double helix one has C^* instead of C in the second position from left in the TCGG... chain, in the newly formed double helix we will have

```
T C* G G        old helix
| |  | |  ...
A A  C C        new helix
```

In this way instead of the original C-G base pair finally (after a second duplication) a T-A base pair will be situated at the same position. Such a change, when a single base pair is substituted by another one, is called a point mutation. It should be pointed out, however, that through the described shift of the position of a hydrogen atom one can interchange A and G or T and C in the same chain. There is, however, no way to interchange A and C or T or G and C or T in the same chain with the help of shifts of the H-atoms. This is the reason why the base substitutions described in Chapter 2 cannot happen in this way (they are so-called non-Watson-Crick-type point mutations).

Turning briefly to the structure of proteins Fig. 4.10 shows a polypeptide chain which has the -(NH)-(C=O)-CRH- peptide unit repeated.

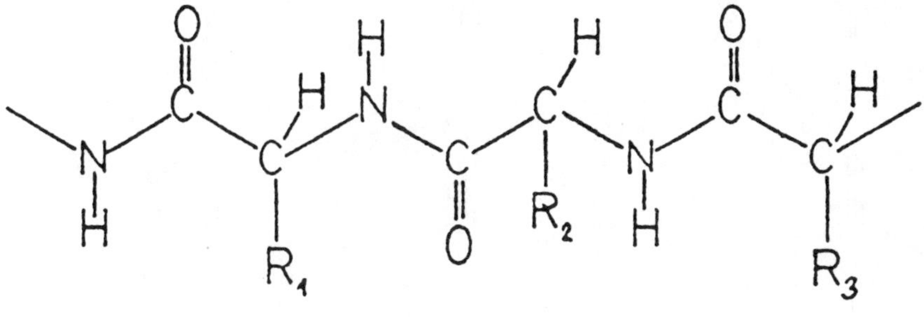

Fig. 4.10: A polypeptide chain with different R_i side chains

Proteins are built of 20 different amino acid units in contrast to DNA which contains only 4 different subunits. Depending on what the side chain R is, we obtain these different amino acid residues. If in the simplest case R = H, we have a glycine unit, if R = CH_3 the unit is called alanine, in the case when R = CH_2-OH serine and so on. Some side chains are rather complicated containing different ring compounds or besides carbon and nitrogen also sulfur atoms. We will not list here all the 20 amino acids but refer to the literature [8]. Due to the fact that proteins have as many as 20 different subunits, the geometrical arrangement of different proteins can be very different and rather complicated. First of all different polypeptide chains are bound together again through hydrogen bonds (see Fig. 4.11).

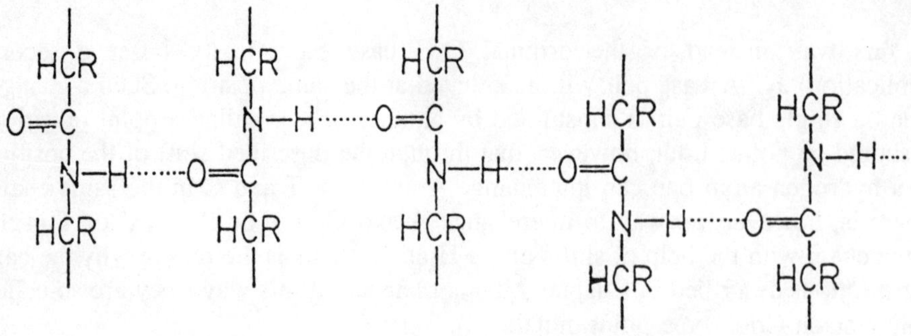

Fig. 4.11: Hydrogen bonds between polypeptide chains

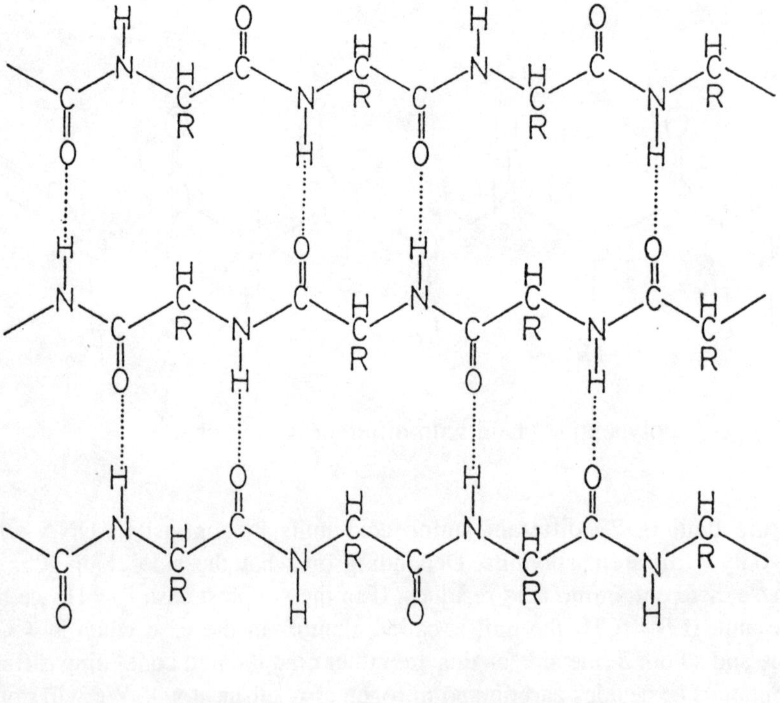

Fig. 4.12: The β-pleated sheet structure of protein. The *dotted lines* indicate hydrogen bonds

These hydrogen bonds can occur between different polypeptide chains, like in the planar antiparallel β-pleated sheet structure of proteins (see Fig. 4.12) or in a three-dimensional structure.

The other possibility is the formation of hydrogen bonds between different parts of the same polypeptide chain in the α-helical structure of protein (see Fig. 4.13). This α-helix is right-handed as the helices in DNA B, and contains 3.6 amino acids per turn. The height of one turn is 5.4 Å, that is we have a rise of about 1.5 Å per amino acid unit in the direction of the main axis of the α-helix. Both the β-pleated sheet and the α-helical structure were first postulated on the basis of the geometrical arrangement of the subunits and have been experimentally verified by the X-ray diffraction technique [9].

A real protein which contains between 50 - 1000 amino acid units has usually a very complicated three-dimensional geometrical structure in which α-helix, β-pleated sheet and more random regions follow each other and the whole chain is folded again in a complicated way (tertiary structure). If one speaks about the chemical and geometrical structure of DNA and proteins, the primary structure means the sequence of the chemical constituents (nucleotide bases, or amino acids, respectively), the secondary structure means the local geometrical arrangement (DNA B or Z, or β-pleated sheet, α-helix, respectively) and the tertiary structure means the folding of the whole double helix or protein.

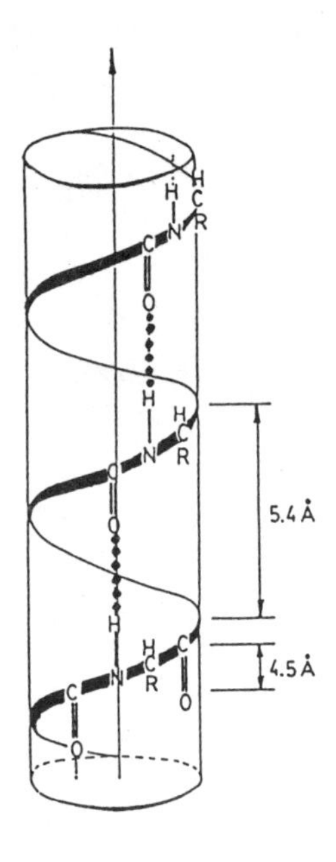

Fig. 4.13: The protein α-helix. The *dotted lines* indicate hydrogen bonds

It should be emphasized that the determination of the complete three-dimensional structure of a complicated protein (finding out the position in space of each of its atomic nuclei) with the aid of X-ray diffraction is a major undertaking and several crystallographers have obtained a Nobel Prize for determining the geometrical structure of an important protein molecule. Fig. 4.14 shows in a schematic way the three-dimensional structure of myoglobin determined by Kendrew as an example [10]. (Myoglobin is a protein which plays an important role in the oxygen metabolism of the cell.) Each circle indicates, in the 153-units-long structure, a complete amino acid unit.

Of course one could write much more about the details of the structure of proteins. Since, however, this is outside the scope of this book here, we would like to mention only the role of disulfur bridges and those of hydrophilic and hydrophobic side groups.

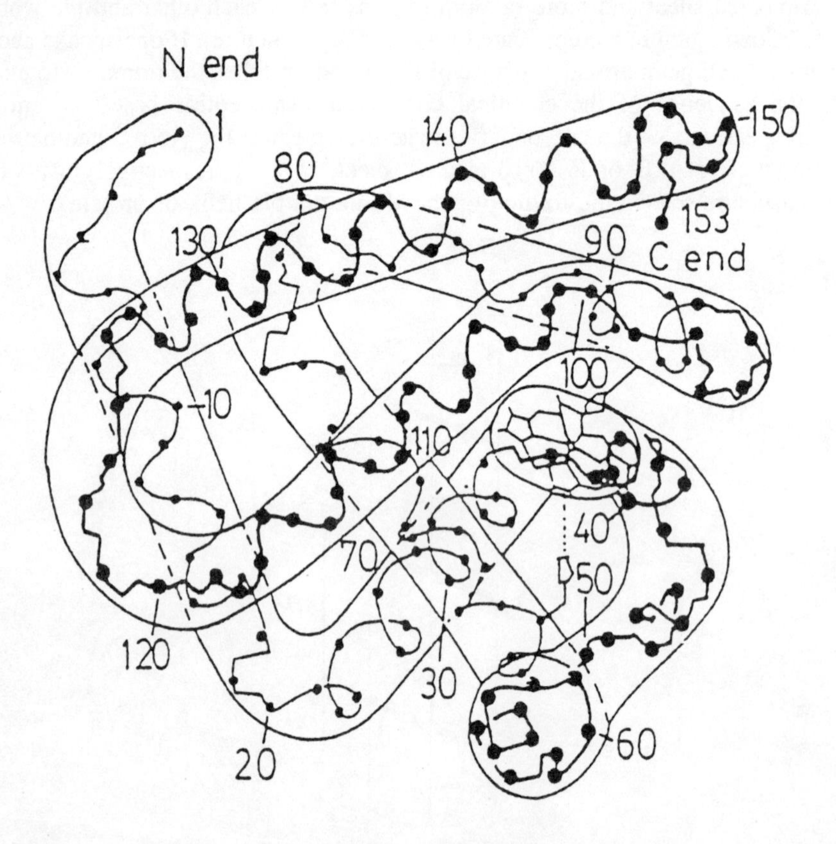

Fig. 4.14: The overall three-dimensional structure of myoglobin. The *circles* indicate the different amino acid units

If one has two cysteine (R = -CH_2-SH) side groups near to each other (not necessarily in the primary sequence, but due to the folding of the whole protein macromolecule) they can form -S-S- bridges between different parts of the main chain. These disulfur bridges contribute significantly to the stability of the actual conformation of the protein molecule in question.

Further if in some region of the protein molecule most amino acid residues have polar side chains, like serine, cysteine, threonine with R = -$CH(CH_3)$-OH, tyrosine in which

R = —CH_2— —OH

etc.), this hydrophilic region will easily interact through H-bonds with water molecules. On the other hand, if in a region there are mostly non-polar side chains (like in alanine, valine with R = -$CH(CH_3)_2$, leucine with R = -CH_2-$CH(CH_3)_2$, etc.) these hydrophobic regions do not bind H_2O molecules, but they can bind well to lipids. The different hydrophilic and hydrophobic regions of a protein are especially important in the case of membrane proteins, because membranes contain lipid-aqueous solution interfaces.

4.2 The Biological Role of DNA and Proteins

Besides duplicating itself (which gives rise finally to cell duplication) DNA of a given cell can recombine with the DNA of another cell in the case of higher organisms. This means that, in the offspring, a part of the genetic information comes from the DNA of one cell and another part from the DNA content of a second one causing the development of a cell (the fertilized egg cell in sexual reproduction) which is different from both parent cells. From this daughter cell through its duplication and differentiation a new organism can develop, different from both parents.

Recently, with the help of new techniques, molecular biologists have been able to slice out genes from the DNA molecules of higher organisms and transplant them into the DNA of bacteria (genetic engineering). Using this, in a certain sense artificial genetic recombination, they transplanted genes into bacteria which produce insulin (this protein is very important for people with diabetes) and interferon (this protein has strong antiviral effects). Very recently the genes which correspond to certain proteins (the so-called antibodies) which bind selectively to certain specific proteins (the so-called antigens) occurring at the outer parts of the cell membranes of tumor cells have been transplanted into bacteria. In this way, these important proteins can be produced in large-quantities in bacteria ("monoclonal" antibodies in bacteria) in a relatively inexpensive way (to synthesize these proteins in larger quantities would be extremely expensive) which opens the door for large scale practical medical applications including the treatment of tumors. Since the description of the ways of genetic recombination lies outside the scope of this book, we cannot discuss this subject here in more detail.

The second main role of DNA both in unicellular and multicellular organisms is to determine the amino acid sequence of the different proteins in the cell. In a human cell e.g. there are more than 5000 different proteins. This happens via another macromolecule, RNA (ribonucleic acid) which has a rather similar chemical but a strongly different geometrical structure to DNA. Protein synthesis occurs in several steps involving three different types of RNA molecules and different protein catalysts. Its site is not the cell nucleus, but it happens at the so-called ribosomes which are smaller particles in the cytoplasm (for further details see [11]).

The sequence of DNA determines the sequence of amino acids in proteins through the so-called genetic code. After many years of extensive experimental work this code has been completely cleared up. The results have shown that the code is universal, i.e. it is independent of the biological source investigated and the location of this source on the earth and it seems to be valid even for bacteria found in meteorites. Three nucleotide bases, a so-called codon, always correspond to one amino acid. It seems that during biological evolution this most effective code was selected. Namely with two bases only $4^2 = 16$ amino acids could have been coded and if four bases coded one amino acid there would be $4^4 = 256$ possibilities causing longer DNA segments to code for a protein in a very highly degenerate manner (very many quartets would code the same amino acid). If on the other hand there were only two different kinds of bases, at least five nucleotide bases ($2^5 = 32$) would be necessary to code one of the 20 different amino acids which would make the DNA segment coding a protein again too long. Finally a code based on the existence of only three different types of nucleotide bases does not come into question, because in this case no double-helical geometrical structure of DNA could have evolved. The code does not overlap. Since there are starting codons which indicate from which position the code has to be to read and from the orientation of the DNA chains one knows also the direction of reading, a non-overlapping code means that the first codon (after the starting signal) corresponds to the first amino acid, the second base triplet to the second amino acid and so on.

Since we have only 20 amino acids in proteins and $4^3 = 64$ base pair triplets, the genetic code has to be degenerate, i.e. more than one base triplet codes the same amino acid). In Ref. [2] Fig. 34-1 (p. 962) gives the complete code and Table 34.3 (on the same page) tells how many different base triplets code the samc amino acid (numbers between 1 and 6). For further details about the genetic code which was one of the greatest discoveries of the last 25 years in molecular biology see besides [2] also [12].

The results about the genetic code described up to now are valid both for unicellular organisms and multicellular organisms (with differentiated cells; see below). Recently [13], however, it has been discovered that in the much more complicated differentiated cells (eukaryotic cells; see Introduction) the parts of the genetic message which code for a protein, the exons (it has been shown previously [14] that to one gene corresponds one protein) can be interrupted by long DNA segments (the so-called introns) which do not code proteins. The biological role of these introns is at the moment unknown. Figure 4.15 shows a DNA segment which contains 5 protein coding regions (exons) and 4 introns between them.

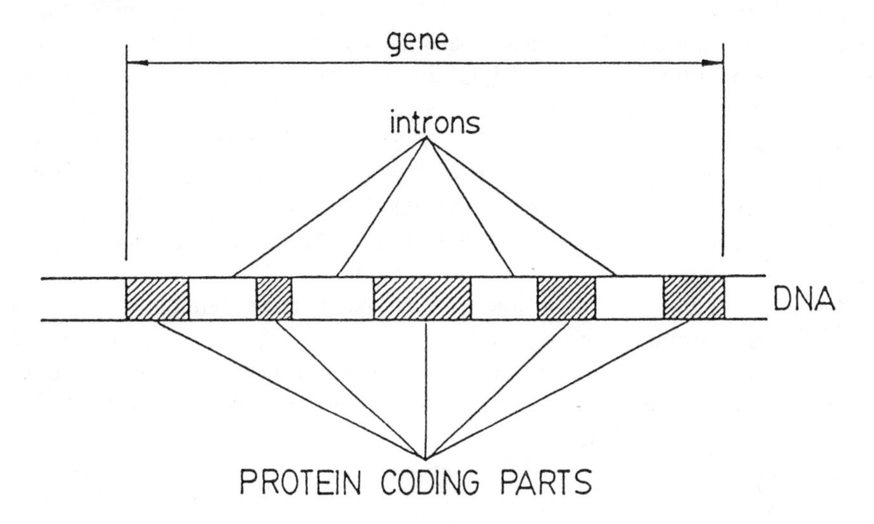

Fig. 4.15: A gene in 5 parts of an eukaryotic cell (schematic)

Proteins have two main functions in the cell: (1) they are very important building stones of higher particles (like the nucleoproteins in the cell nucleus, proteins in the membranes, ribosomes, mitochondria etc.) in the cells (structural proteins) and (2) they are the catalysts of biochemical reaction in them (enzymes).

As mentioned in the Introduction most of the about 10^5 genes of an eukaryotic cell of a higher organism are permanently blocked (in a human cell there are in average somewhat more than 5000 different types of proteins, so for their coding only somewhat more than 5000 genes are needed). As established experimentally [15], every cell in our body (or any differentiated cell of any higher organism) contains exactly the same total genetic information. The fact that which of the about 5000 genes out of about 10^5 are actively coding proteins decides what kind of cell we have: a skin cell, a liver cell, an eye cell, a lung cell and so on. The non-active majority of the genes is kept inactive by proteins which bind to them and in this way block them. In other words we can say in a somewhat oversimplified way that the regulation of a differentiated (eukaryotic) cell depends on DNA-protein interactions and by influencing DNA-protein interactions we can influence genetic regulation.

The other main role of proteins is that they serve as catalysts of the about $5x10^4$ biochemical reactions going on in a cell of a higher organism. One protein can perform enzymatic (catalytic) activity for several or in some cases for a larger number of reactions. Since biochemical reactions are very fast though they take place at room-temperature, we can say that the protein catalysts (enzymes) speed them up to a great degree (with a factor between 10 and 10^7). The mode of action of the enzymes is still the subject of intensive research, but one can say already that the catalytic activity of an enzyme is dependent only on a small part of the protein molecule (the so-called active site). The rest of the protein molecule has - according to all probability - the role of keeping the active site in the right conformation, so that the reacting molecules can bind to it easily, and provides it with the right electron distribution which also can facilitate a reaction to a great degree.

From the foregoing discussion we can see that the genetic information carried by DNA determines (through RNA) the structure of the proteins occurring in the cell which in their turn determine which chemical reactions will take place. In this way DNA (which we could consider analogous to the program of a computer) in the final analysis determines through the proteins (which we can visualize, to continue the analogy to a computer, as its processor) the structure and biological function of a whole cell. We can express symbolically this situation as: DNA $\rightarrow$ RNA $\rightarrow$ P $\rightarrow$ biochemical reactions $\rightarrow$ structure and function of the cell (here P stands for protein).

References

1. O.T.C. Avery, G.M. McLeod and M. McCarty, J. Exp. Med. **79**, 137 (1944).
2. See for instance: A.L. Lehninger, Biochemistry, Worth Publ. Inc., New York, 1975, Chapter 12.
3. M.H.F. Wilkins, A.R. Stokes and H.R. Wilson, Nature **171**, 738 (1953); M.H.F. Wilkins, W.E. Seeds, A.R. Stokes and H.R. Wilson, Nature **172**, 759 (1953):

4. J.D. Watson and F.C. Crick, Nature **171**, 737, 964 (1953); F.H.C. Crick and J.D. Watson, Proc. Roy. Soc. (London) **A223**, 80 (1954), J.D. Watson, The Double Helix, Atheneum Press, New York, 1968.
5. See for instance: R.A. Dickerson, H.R. Drew, B.N. Conner, R.M. Wing, A.V. Fratini and M.L. Kopka, Science **216**, 475 (1982).
6. A.H. Wang, G.J. Quigley, F.J. Kolpak, J.L. Crawford, J.H. van Boom, G. van Mareland, A. Rich, Nature **282**, 680 (1973); A.H. Wang, G.J. Quigley, F.J. Kolpak, G. van der Marel, J.H. van Boom and A. Rich, Science **211**, 171 (1980).
7. M. Meselson and F.W. Stahl, Proc. Math. Ac. Sci. USA **44**, 671 (1958); A. Kornberg: DNA Synthesis, Freeman Inc., San Francisco, 1974.
8. See for instance [2] or Table 1-1 (p. 6) in S.A. Bernhard: The Structure and Function of Enzymes, W.A. Benjamin, Inc., Menlo Park. Ca.-Reading, Mass.-London-Amsterdam, 1968.
9. L. Pauling, R.B. Corey and H.R. Branson, Proc. Nat. Acad. Sci. USA **37**, 205 (1951); see also [8].
10. J.C. Kendrew, Science **139**, 1259 (1963) (Basically the text of Kendrew's Nobel Lecture).
11. P.H.W. Butterworth and T.J.C. Beebe in Biochemistry of Cellular Regulation, M.J. Clemens ed., CRC Press Inc. Boca Raton, Fla. Vol. 1 (1980) p.1.
12. M.W. Nirenberg and J.H. Matthaei, Proc. Nat. Acad. Sci. USA **47**, 1588 (1961); M.W. Nirenberg and P. Leder, Science **145**, 1399 (1964); F.H.C. Crick, J. Mol. Biol. **19**, 548 (1966); M. Ycas, The Biological Code, North Holland, New York, 1969.
13. See for instance: W. Gilbert, Nature **271**, 501 (1978).
14. G.W. Beadle and E.L. Tatum, Proc. Nat. Ac. Sci. USA **27**, 499 (1941).

5 Mechanism of Oncogene Activation or Antioncogene Inactivation by External Factors

5.1 Carcinogens and Radiation Affect DNA Not Only Locally

As briefly described in Chapter 1, Busch [1] was the first to postulate the existence of cancer-causing genes in normal cells of higher organisms. Though he did not write explicitly about human oncogenes, he assumed that due to carcinogens binding to proteins, these cancer-causing genes become readable by releasing the proteins which block them and so their information content can initiate cancer in a cell. Nowadays we know that - according to all probability - carcinogens exert their effect first of all through binding to DNA. In a paper, one of the present authors (J.L.) together with Suhai and Seel [2], assuming the existence of human oncogenes, reviewed the different possible local and long-range effects through which a chemical carcinogen bound to DNA can interfere with DNA-protein interactions and in this way with the genetic regulation of the oncogenes. Fig. 5.1 shows this so-called reading error theory of Busch modified in that it is assumed that the carcinogen (C) binds to DNA, but not necessarily only at the site of an oncogene.

A simple consideration shows that even if the carcinogens bind at the site of an oncogene, alone through local effects it is extremely improbable that they can cause the release of a protein (P) from the DNA-P complex and in this way deblock an oncogene. One knows that nucleohistone is the protein which most frequently blocks a gene in DNA-P complexes. This protein is rather rich in the amino acid arginine (see Fig. 5.2). The arginine molecule has as end group, a so-called guanidium group which can form a hydrogen-bonded complex with the phosphate group of DNA (see Fig. 5.3). Therefore, one can model the interaction of an arginine side chain of a polypeptide chain with DNA through this guanidium-phosphate complex. Actually this complex also exists in a crystal form and so an X-ray investigation by Cotton and his coworkers [3] was able to provide an accurate geometrical structure for it. Knowing the position of its atomic nuclei it was not difficult to perform a quantum theoretical calculation of the distribution of its electrons providing a binding energy of about 2.5 electronvolt (eV) = 57.5 kcal/mol (1 eV = 23 kcal/mol) essentially of electrostatic nature [4]. To describe the distribution and behaviour of electrons in an atom, molecule, macromolecule or solid modern theory applies quantum mechanics to take into account the basic fact that electrons in these systems can have only well-defined, discrete energy values, i.e. their energies are quantized. The above mentioned value of 2.5 eV is a rather large interaction energy which is similar to the binding energies of some normal chemical bonds. In reality, this binding energy can be much smaller due to the effect of ions which interfere with the electrostatic binding in this molecular complex (screening). However, most probably it is still essentially larger than the hydrogen bond energies of the complex.

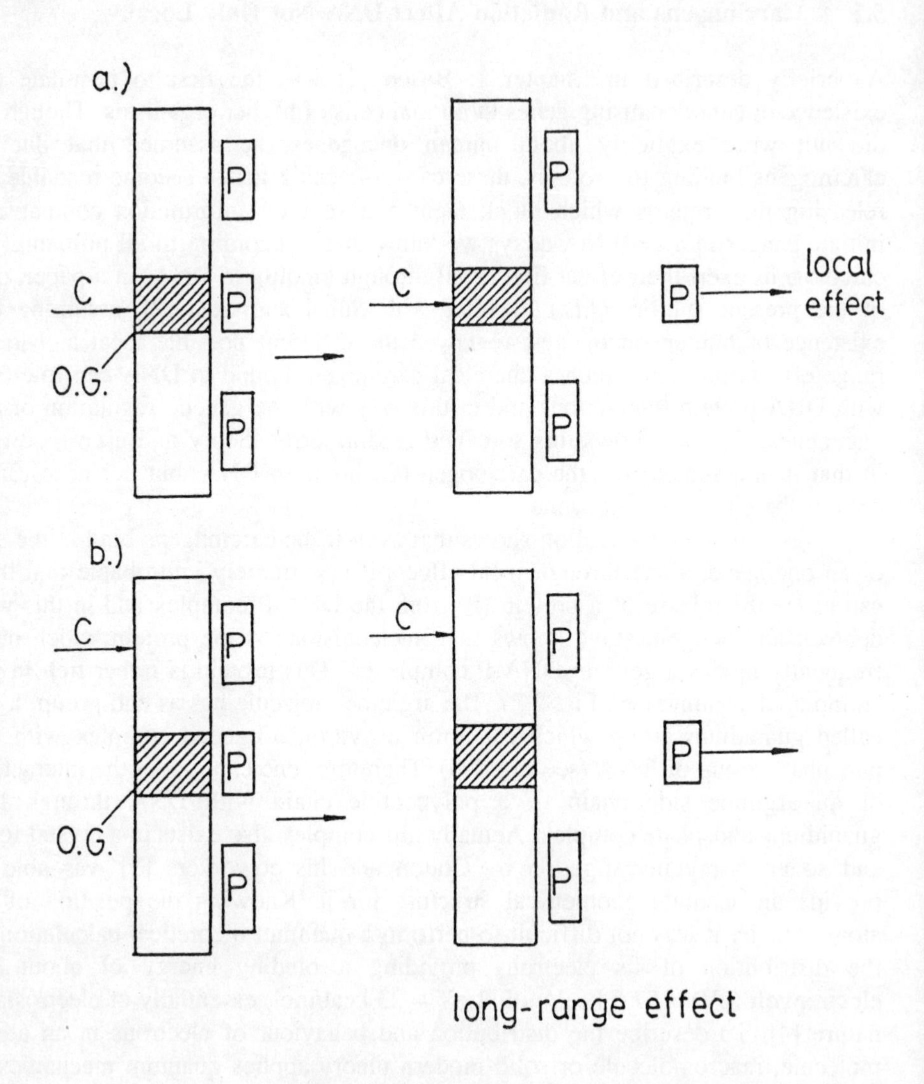

Fig. 5.1: Representation of the reading error theory of chemical carcinogens. In part (a) of the Fig. the carcinogen (*C*) is binding to an oncogene, in its part (b) to another part of DNA which doesn't belong anymore to the oncogene. *P*. stands for protein

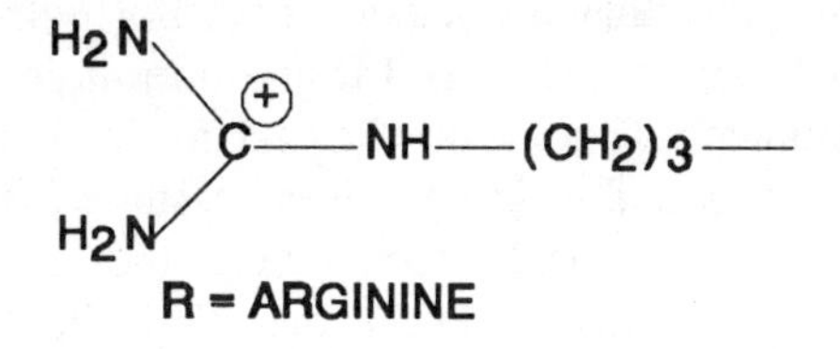

Fig. 5.2: The chemical formula of arginine

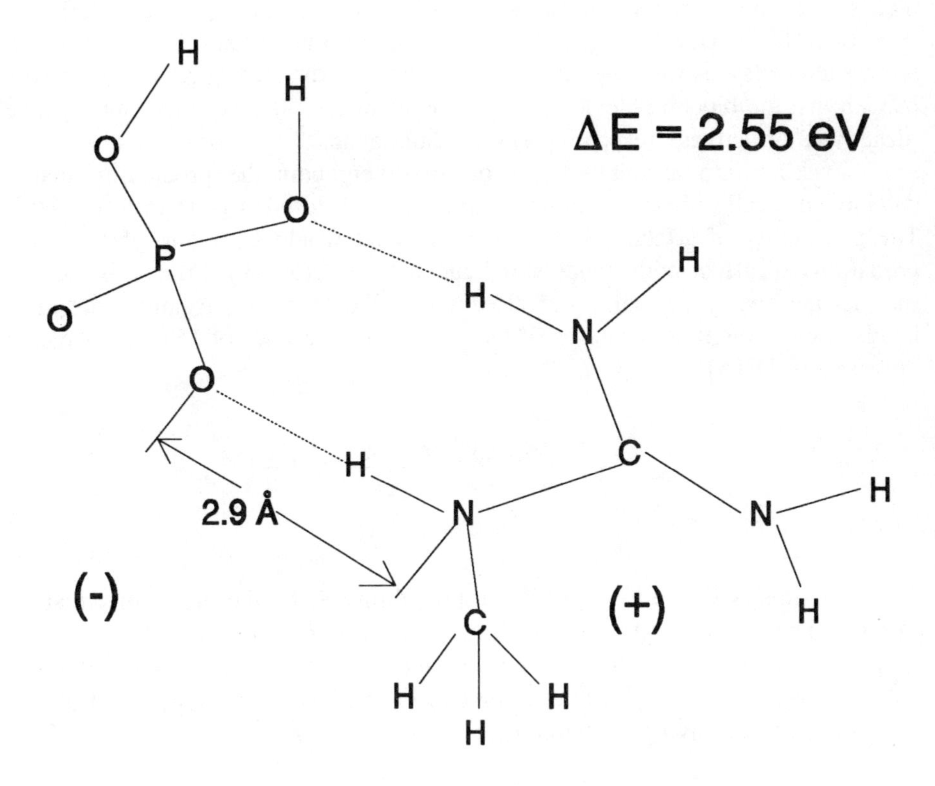

Fig. 5.3: The chemical structure of the phosphate-guanidium complex. ΔE stands for the interaction energy

Let us assume now that a nucleohistone molecule which is 300 amino acids long, has only 5 per cent, that is 15, arginine molecules in it (which is a very conservative estimate). This means that our nucleohistone molecule is at 15 different places strongly bound to DNA. Let us assume further that a carcinogen bound to DNA can have only local effects (say at the base it is bound to and at its first right and left neighbor). From the pitch height of a turn of the nucleohistone (3.6 Å; see in Sect. 4.1) as well from the height of a turn of DNA B (34 Å) and the number of base pairs in its turn (10) one can easily calculate that about 20 (more precisely 19) amino acids correspond to 10 stacked bases in the DNA chain (if we assume an α-helix configuration for the protein chain). This means that a nucleohistone of 300 amino acids can block a segment of 150 base pairs of DNA B. Usually a gene is on average 1000 units long, but since in differentiated cells - as we have seen - it can consist of different pieces, it can easily happen that the blocking-deblocking of a segment of 150 bases of it can regulate (deactivate or activate, respectively) the whole gene.

Taking into account effects on first neighbors the probability that a carcinogen locally affects a strong arginine-phosphate bond is 1/(150/3) = 1/50. The probability of affecting a second such bond would be again 1/50 and the probability to affect both bonds simultaneously is 1/(50•49). From this we can see that the overall probability of affecting all the 15 strong arginine-phosphate bonds at the same time would be (if we assume the presence of 15 carcinogens in this exon of DNA)

$$P_1 = \frac{15}{50} \cdot \frac{14}{49} \cdot \frac{13}{48} \cdots \frac{2}{37} \cdot \frac{1}{36} \ll \left(\frac{1}{3}\right)^{15}$$

Further we have to multiply the probability P_1 by the probability that 15 carcinogens get at all to the 150 base pair long exon of DNA under consideration. It is well known that an animal has to be injected with a concentration of 10^{-5} - 10^{-7} carcinogen per base pair to develop cancer [5]. Taking the upper limit of 10^{-5} the probability of having 15 carcinogens at the 150 base pair long exon is [6]

$$P_2 = \left(\frac{150}{15}\right)^{15} = 1.5^{15} \cdot \frac{(10^2)^{15}}{10^{75}} = 1.5^{15} \cdot 10^{-45}$$

In this way the joint probability that one breaks the 15 strong bonds at the 150 base pair long regulatory part of an oncogene is [6]

$$P = P_1P_2 = \frac{15}{50}\cdot\frac{14}{49}\cdots\frac{1}{36}\cdot 1.5^{15}\cdot 10^{-45} \ll \left(\frac{1}{3}\right)^{15}\cdot 1.5^{15}\cdot 10^{-45} =$$

$$= \left(\frac{1}{2}\right)^{15}\cdot 10^{-45} \approx 3\cdot 10^{-50}$$

which is an extremely small number.

From this simple estimation we can conclude that to deblock even a part of an oncogene is extremely improbable with the aid of carcinogens bound to it, if we assume only local effects of them. On the other hand, as we have seen in Chapter 2, with the exception of oncogene-activation through point mutation, all the other ways they are activated presume their deblocking from their protein shield. The only way out of this dilemma seems to be that we assume that carcinogens bound to DNA also have long-range (usually solid state physical) effects. Before describing the different probable long-range effects of carcinogens we shall in the next point summarize their local effects (which of course still play a role). After that, we will sketch in a qualitative way, how one can describe the behavior of electrons in a macromolecule - like DNA or proteins - if one treats these electrons from the point of view of solid state physics.

Finally, one should point out that the same considerations about long-range effects refer not only to chemical carcinogens, but also to electromagnetic (ultraviolet and X-ray) and to particle radiation. If the radiation hits DNA directly, one cannot explain well the release of a protein from DNA or double strand breakings assuming only local effects. In the case of high energy radiation, if the energy taken from the radiation by DNA is relatively large, most probably this will not lead to the deblocking of an oncogene, but will have either lethal effects or lead to inactivation of antioncogenes, for instance by breaking both strands of DNA leading to a substantial loss or rearrangement of the genetic information). Low energy radiation (if it hits DNA directly) can again cause the release of nucleohistones and therefore the activation of oncogenes, if also long-range effects along the DNA chains are taken into account.

In most cases, radiation does not hit DNA directly, but its energy is absorbed in the cytoplasm causing the formation of radicals, which are very reactive substances, because they possess an odd number of electrons. These radicals can then bind to DNA and therefore all the above-described considerations about chemical carcinogens binding to DNA are also valid for them.

5.2 Local Effects of Carcinogens and Their Effect on Oncogene Activation (Point Mutations)

A bulky chemical carcinogen bound to a DNA constituent first of all can distort the geometrical structure of the double helix. For instance the ultimate (final reaction product inside the cell) of the prominent carcinogen

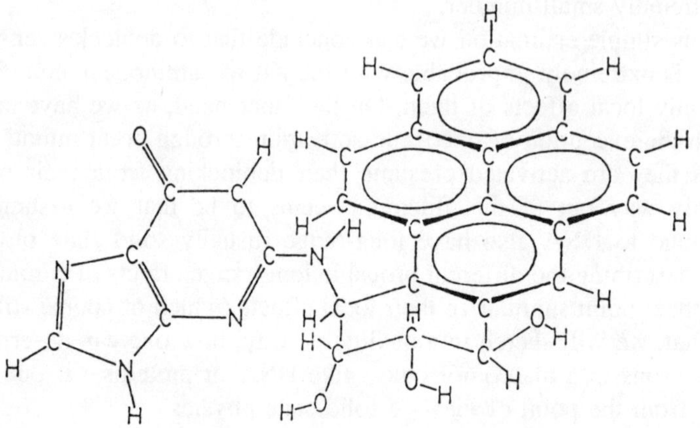

Fig. 5.4: The ultimate of 3,4-benzpyrene bound to the amino group of guanine

3,4-benzpyrene (see Chapter 3) which binds to the -NH_2 group of G (see Fig. 5.4) [7] certainly distorts the local conformation of DNA. As another example we can take the compound AAF (see Chapter 3) which binds to the carbon atom of the five-membered ring of G. In this case AAF usually takes the place of G in the stack of nucleotide bases in DNA B and the guanine molecule sticks out from the double helix (see Fig. 5.5) [8]. On the other hand, if the guanine molecule to which AAF is binding, is a part of an alternating GCGCGC... sequence the originally B conformation of this part of DNA can easily go over to a Z conformation [8].

Another local effect of carcinogen binding to DNA can be a charge redistribution of which charge transfer is the most important special case. There is a charge transfer between two interacting molecules A and B, if a net charge is flowing over from one molecule to the other. We can symbolically write for this process $A^{\delta+}$-$B^{\delta-}$ in which case A is the electron donor (has given charge) and

B is the electron acceptor (it has received electric charge). In the case of the charge transfer complex $A^{\delta-}$-$B^{\delta+}$ the roles of A and B, respectively, are interchanged. Of course the electron attracting power (the so-called electron affinity) determines which one of the molecules A and B will be the acceptor and which one the donor. As we shall see in the next point charge transfer can also cause long-range effects along a DNA chain.

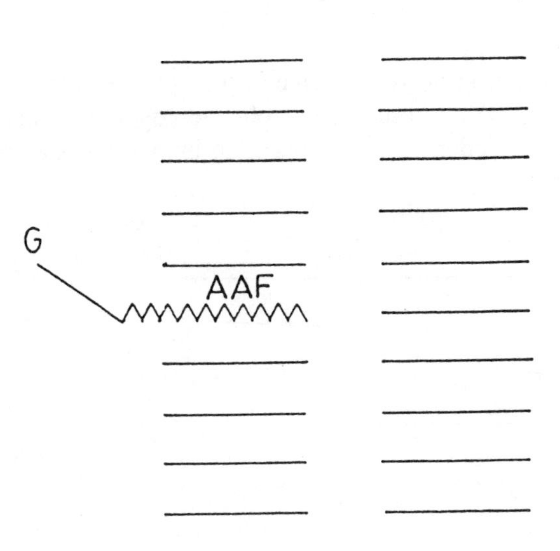

Fig. 5.5: AAF takes the place of a G molecule in the base stack and the G molecule sticks outside

One should point out that charge redistribution (charge transfer) can easily cause such a shift of hydrogen atoms (see Fig. 4.2) leading to a Watson-Crick-type point mutation (see Sect. 4.1). On the other hand, first of all, a distortion of the double helix is needed to achieve for instance a

$$\begin{matrix} G & & G \\ G & \rightarrow & T \\ C & & C \end{matrix}$$

type non-Watson-Crick-type point mutation (as mentioned already in Chapter 2) in which in the same chain a bulky (A or G) nucleotide base is exchanged to a smaller one (T or C).

On the other hand, if we assume a local distortion of the double helix due to carcinogen binding (see Fig. 5.6), one can assume that in the distorted helix there is enough room for the unusual G-A base pair (see Fig. 5.7). In this way we can postulate the following sequence of events

$$
\begin{matrix} G & & C & & G \\ G & \rightarrow & A & \rightarrow & T \\ C & & G & & C \end{matrix}
$$

If in the course of replication an A molecule is incorporated instead of C in the complementary strand of DNA, in the next duplication cycle we obtain instead of the triplet CGG the CTG codon as it was found in human EJ bladder carcinoma ([9]; see also Chapter 2).

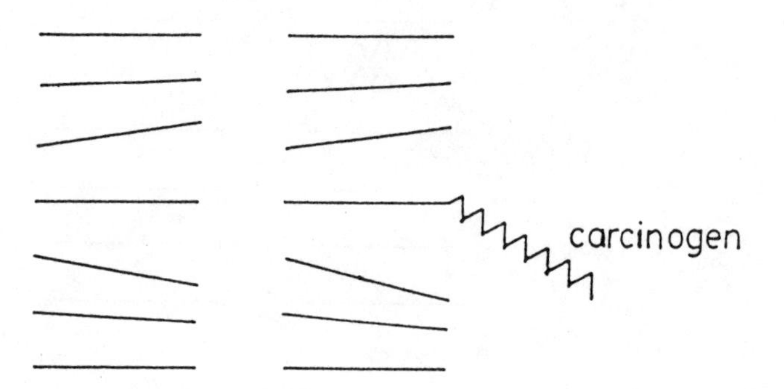

Fig. 5.6: The local distortion of a nucleotide base pair stack due to the binding of a carcinogen (schematic)

Carcinogens bound to DNA can, of course, also affect the tertiary structure of DNA at the site of their binding. As we shall see in the next point this may again cause long-range effects.

If CH_3- or CH_3-CH_2- groups are bound to the nucleotide bases due to the effect of alkylating agents (see Chapter 3) they usually bring a whole positive elementary charge to these molecules. In this way they can interfere with the base pairing causing point mutations and therefore have a carcinogenic effect. On the other hand, the positive charges brought by these alkyl groups (see Chapter 3), even if these groups do not bind to atoms which take part in hydrogen bonds, can strongly influence the electronic charge distribution of the nucleotide bases and in this way (similarly to the charge transfer) can have long-range effects.

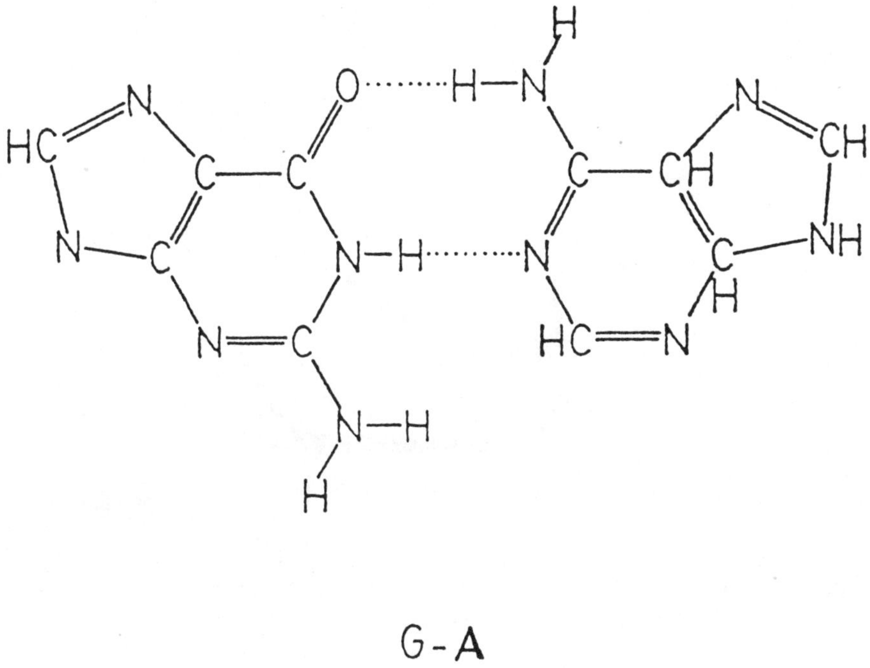

G-A

Fig. 5.7: The unusual G-A base pair

The binding of carcinogens (especially if they are larger molecules) to DNA can strongly influence locally also the vibrations of the DNA constituents. In a molecule the nuclei of the atoms are not at rest, but they move around their equilibrium positions in a periodic way. These periodic motions of the atoms are called vibrations. The local change of vibrations can have, however, an influence on the conduction properties of the DNA chains and in this way they can exert again a long-range effect (see [2]).

Finally, as it is well known, carcinogen binding or radiation effects can cause the breakage of chemical bonds and the formation of new bonds. A chemical example of this is the dimerization of two stacked pyrimidine molecules due to ultraviolet radiation [10] which of course is again accompanied by the distortion of the original geometrical arrangement of the double helix of DNA B (see Fig. 5.8).

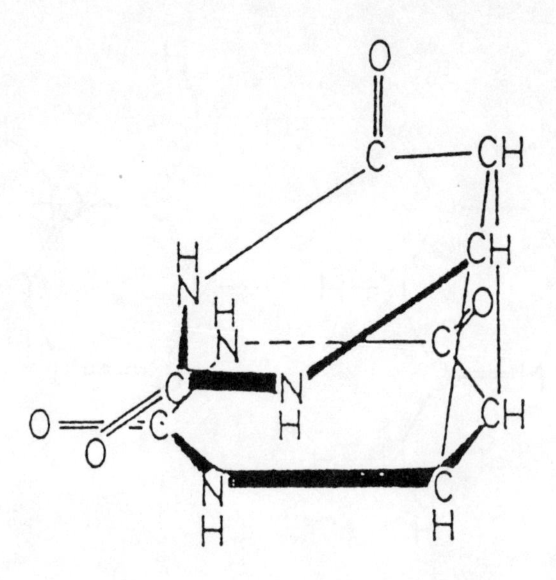

Fig. 5.8: The dimerization of two stacked pyrimidine molecules due to ultraviolet radiation (schematic)

One should emphasize that though we have concentrated here on the local effects of carcinogen-binding to DNA, the same effects can occur (probably with the exception of point mutations) if the carcinogen is bound to the protein part of a nucleoprotein influencing in this way the DNA-P interactions. For a review of local effects of chemical carcinogens bound to DNA see [2].

5.3 Different Long Range Physical Effects of Carcinogens

5.3.1 Energy Bands in DNA and Proteins and Their Change Through Carcinogen Binding

To understand the different long-range effects of carcinogens which are based on solid state physical phenomena, one has to discuss first the electronic structure of DNA and proteins from a solid state physical point of view. According to the modern physics of matter - as mentioned above - the electrons of an atom, a molecule or a solid (including biologically active macromolecules) cannot have arbitrary energies, but only well-defined discrete energy levels. If we have only one electron, as in the hydrogen atom, we can determine these allowed energy levels theoretically with the help of quantum theory in complete agreement with experimental (spectroscopical) results. On the other hand, if the number of electrons is larger than one, the problem cannot be solved exactly, but in the last five decades rather accurate approximate methods have been worked out for the determination of the energy values of a system with many electrons including the calculation of the electronic distribution in them. This kind of research which is the main subject of theoretical atomic physics, molecular physics which is also called quantum chemistry and solid state physics is very important because the energy levels and the distribution of the electrons in these systems determine completely their different physical and chemical properties.

If we have a molecule say with 4 energy levels, each level will be doubled if two such molecules interact. If the number of identical molecules is three, each level will be tripled and so on. Finally, if the number of interacting molecules is very large (in mathematical language it approaches infinity) which happens in a crystal or in a long chain, the very large number of close lying energy levels merges into a continuous region of allowed energies of the system, i.e. a so-called energy band is formed. Fig. 5.9 shows this situation.

The highest filled energy band of a crystal or long chain is called the valence band and the lowest unfilled one the conduction band. The energy difference between the upper edge of the valence band and the lower limit of the conduction band is referred to as the fundamental gap. If one speaks about all the energy bands of a solid or a chain, this is called the energy band structure of the system. Detailed considerations show that if the valence band of a system is completely filled and the gap is large, no electric conduction can take place, the system is an insulator. The thermal energy can now excite higher vibrational states or can cause collisions in a solid. Note, that the energy of such vibrations is not arbitrary but again quantized. If the valence band is now filled, but the gap is rather small and if we are not at absolute zero temperature, this thermal energy can excite a number of electrons from the valence band to the conduction band, and therefore a not too strong current can flow through the solid. Such a solid is called a semiconductor. Finally if the valence band is not completely filled, or the conduction band is partially filled, strong electric conduction can take place (the system is a conductor). In this case, primarily, the widths of the partially filled bands and the strength of the interaction of the electrons in these bands with the vibrations of the system determine how good a conductor is (how well does it transport electric charge). Metals are usually good conductors, because besides having partially filled valence or conduction bands these bands are rather broad. On the other hand in the last decade molecular chains have also been discovered which conduct as well as a good metal. Figure 5.10 shows in a schematic way the valence and conduction bands of an insulator, of a semiconductor and of a conductor, respectively. For an introduction to theoretical solid state physicsone can recommend Kittel's book [11], for a review about the electronic structure

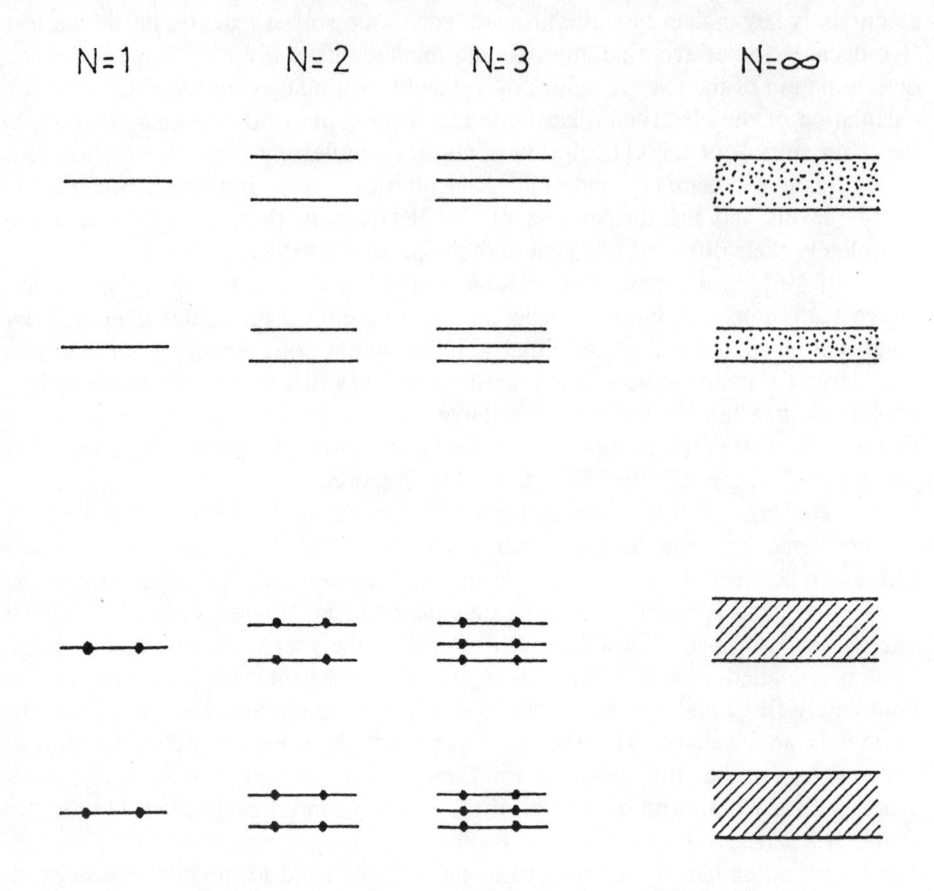

Fig. 5.9: Four energy levels of a molecule of which two are filled (each one by two electrons indicated by *dots*) and two are unfilled. If two such molecules interact each level will be doubled, if there are three identical molecules the original levels of the molecules will be tripled. Finally if the number of interacting molecules is very large, ($N \rightarrow \infty$) continuous energy bands are formed. Here N stands for the number of molecules. In this way two completely filled and two unfilled bands will be formed

(energy bands and distribution of electrons) of highly conducting molecular chains see [12].

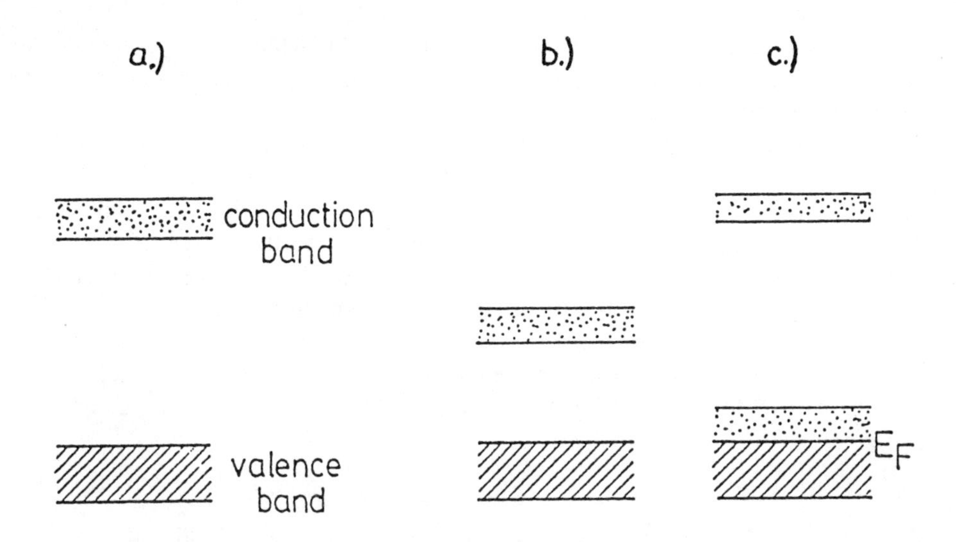

Fig. 5.10: The valence and conduction bands of (a) an insulator (b) of a semiconductor and (c) of a conductor. (E_F stands for the Fermi level.)

In the case of DNA the sugar-phosphate backbone is always periodic (see Fig. 4.1) and therefore it has a band structure. The detailed [13] calculations have shown, however, that the valence band of this chain is rather narrow (about 0.10 eV), but its conduction band has a width of about 0.3 eV. The nucleotide base stacks in DNA are of course aperiodic. In a non-periodic (disordered) system the energy band structure is gradually destroyed with increasing disorder and one can speak instead of energy bands only about the distribution of energy levels. However, to obtain a first orientation about the energy level distribution in the base stacks of DNA, one can perform calculations of the energy bands of periodic polyA, polyT, polyG and polyC. These so-called periodic nucleotide base model stacks have to be understood such, that we have always the same nucleotide base repeated in the same relative position as in DNA B (3.36 Å stacking distance, 36° rotation). One should point out, however, that on the one hand the statistical analysis of the base sequences of many DNA fragments has shown that there is a preference in native DNA to have the same base repeated several times (in a few cases many times, up to 30) [14]. On the other hand all these so-called homopolynucleotides (a sugar-phosphate chain to which always the same nucleotide base is bound) have been synthesized in the laboratory and they are commercially available.

Table 5.1 shows the widths of the valence and conduction bands of all the four base stacks in the DNA B configuration [13]. The relative positions of these bands can be seen from Fig. 5.11. As it is shown in the Table the band widths vary between 0.3 and 0.8 eV. They are very similar to the corresponding band widths of a highly conducting molecular crystal, the so-called TCNQ (tetracyanoquinone) -TTF (tetrathiafulvalene) system. In the latter case the band widths are between 0.3 and 1.2 eV if one performs the calculation exactly in the same way [15]. In this system there are free charge carriers (partially filled valence and conduction bands, respectively) due to a large internal charge transfer between

Table 5.1: The widths of the valence and conduction bands of the four nucleotide base stacks in the DNA B configuration (in eV-s) [13]

	Valence Band	Conduction Band
Adenine Stack	0.45	0.31
Thymine Stack	0.61	0.32
Guanine Stack	0.83	0.75
Cytosine Stack	0.87	0.82

the two different molecules which form alternating stacks. This fact suggests the idea that if one could generate free charge carriers in periodic DNA stacks (homopolynucleotides) by interaction with organic electron acceptors or donors, periodic DNA would also become a good conductor, although it is in itself an insulator due to the large gap (see Fig. 5.11). The interacting organic molecules in this case would take out electrons from the valence band or would put electrons into the empty conduction band, respectively, of the DNA stacks. This experiment is, of course, not an easy one, because before homopolynucleotides can be doped, i.e. brought into interaction with electron acceptors or donors, they have to be purified and characterized to a great extent. This material science development, without which no serious solid physical measurement could be done on DNA or proteins (see also Sect. 7.6) is, however, very expensive. Anyway there is a well-established hope that within a few years the experiment will be possible and we shall have highly conducting periodic DNA.

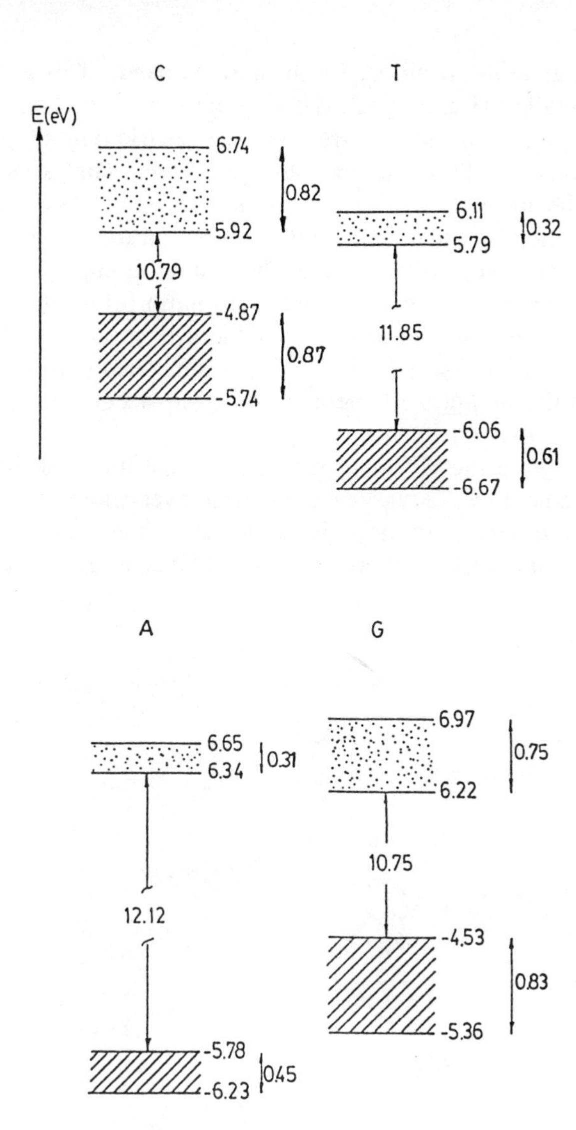

Fig. 5.11: The valence and conduction bands of the four nucleotide base stacks

In connection with the band structures of the base stacks one should point out that the method used for their calculation always gives a too large gap. This calculation method gives a comparatively realistic position for the filled bands (including the valence band) at least on a relative scale, but always provides incorrect (too high lying) positions for the unfilled bands. This shortcoming of the method, the so-called Hartree-Fock (HF) crystal orbital method [15] stems from the basic assumption that each electron sees the electric field of all the nuclei and a suitably averaged field of all the other electrons. This assumption is quite reasonable in the lowest energy state (ground state) of a system, but becomes invalid if we excite electrons. This is the main reason for the gap being too large. There are, however, well established methods how to correct this failure of the HF method, if one works with its so-called ab initio form [16], where one takes into account explicitly all the electrons of the system and all their interactions within the framework of the HF theory. The results cited until now were all done with the help of this ab initio HF method taking into account the periodic (helical) symmetry of the DNA chain.

To go beyond the HF method to correct the gap, one has to apply still more refined methods. Since such calculations are very time consuming they have been executed until now only for periodic chains with an elementary cell containing only a few atoms such as alternating *trans*-polyacetylene (see Fig. 5.12 [17]).

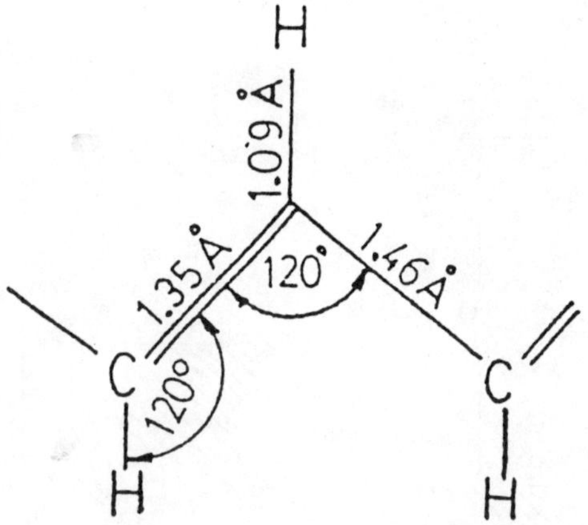

Fig. 5.12: The geometrical structure of alternating *trans*-polyacetylene

The results of these refined calculations have shown quite good agreement with experiment for the gap. The same kind of refined calculation has also been partially performed for a C stack [18]. From the results obtained one can extrapolate that the fundamental gap in periodic nucleotide base stacks is around 5 eV.

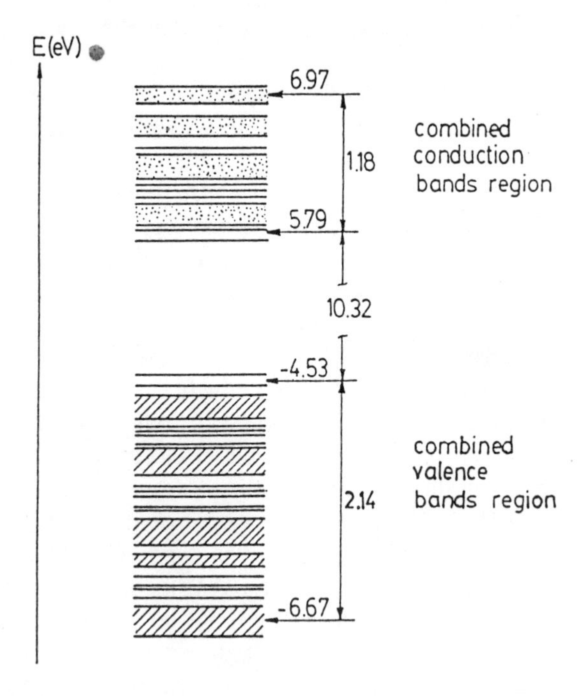

Fig. 5.13: The effect of disorder on the valence and conduction band regions of the nucleotide base stacks

Real DNA is of course aperiodic, but as appropriate calculations show (which we cannot discuss here due to the complexity of the methods), the energy levels of non-periodic DNA are situated in the same regions which are determined by the superposition of the energy bands of the four periodic base stacks [19] (see Fig. 5.13). The main effect of aperiodicity is that even if aperiodic DNA were doped to generate free charge carriers, and the Fermi level (that energy level until which the levels are filled) fell into a continuous region of allowed energies, it would still not be certain that the DNA stacks would have a larger conductivity. Namely, according to the theory of disordered systems most of the electrons become localized (move only in a small part of the system) and therefore the conductivity decreases in a great degree (for this so-called Anderson localization see [20]). The question of conductivity in an aperiodic chain is by no means a trivial one, and a large number of experimental and theoretical investigations are addressed to this problem all over the world. It should be mentioned, however, that very recently the conductivity by hopping in a native disordered protein (pig insulin A) has been calculated using quite complicated theoretical methods [21].

The results have shown that this disordered protein has a conductivity comparable to that of semiconducting disordered glasses [21,22].

Finally if one speaks about the band structure of DNA, one should not forget that DNA or a nucleoprotein is not in a vacuum, but is surrounded by quite structured water and K^+ ions. Clementi and his coworkers have performed very extensive studies on this problem both for DNA B and Z DNA assuming different sequences of the bases [23]. Using his water structure around a C-sugar-phosphate unit, the energy bands of a C stack were recalculated in Erlangen. The results show roughly a 10% broadening of the bands and a downward shift of about 1 eV of their positions [24]. Though these changes are not dramatic, they are not negligible. In a subsequent calculation on the effect of the ions around DNA (which is in progress), one would expect much larger changes.

If a carcinogen binds to DNA, either a bulky one or an alkyl group with one positive elementary charge, this introduces further disorder in DNA B. The sugar-phosphate chain, though it is periodic, sees a non-periodic electric field due to the aperiodic base stacks. This non-periodic perturbation will be strengthened if foreign molecules bind to the bases in a random fashion, especially if they introduce net electric charges randomly. As a consequence of this, the band structure of the sugar-phosphate chain can be destroyed partially or fully, depending on the frequency with which carcinogens bind to DNA. This will certainly decrease the conductivity in this chain, even if free charge carriers are present in it. In addition random carcinogen binding will decrease further the possibility of conduction along the DNA stack. Finally, introducing new gaps into the bands of the sugar-phosphate chain, the carcinogens will most probably decrease the important attractive dispersion (mutual polarization) interactions between the sugar-phosphate chain and a polypeptide chain (in more detail in the next point).

The electronic structure of proteins, due to their 20 constituents and to their different conformations is much more complicated than that of DNA. In Fig. 4.10 the chemical and geometrical structure of the backbone of a polypeptide chain was sketched. On the other hand strong hydrogen bonds exist between adjacent polypeptide chains in the β-pleated sheet structure of proteins. Thus for the calculation of the electronic structure of such a protein, one has to consider it as being at least a two-dimensional system (see Fig. 5.14). Early simpler calculations [15] which were performed on this two-dimensional network have indicated that the resulting band structure is different from the superposition of the two one-dimensional band structures. The corresponding ab initio band structure calculations on polyglycine were performed with one-dimensional models taking separately the main chain and the hydrogen-bonded chain [2] into account. In Table 5.2 the widths of the valence and conduction bands of polyglycine, calculated along the main chain or along the perpendicular hydrogen-bonded chain are presented.

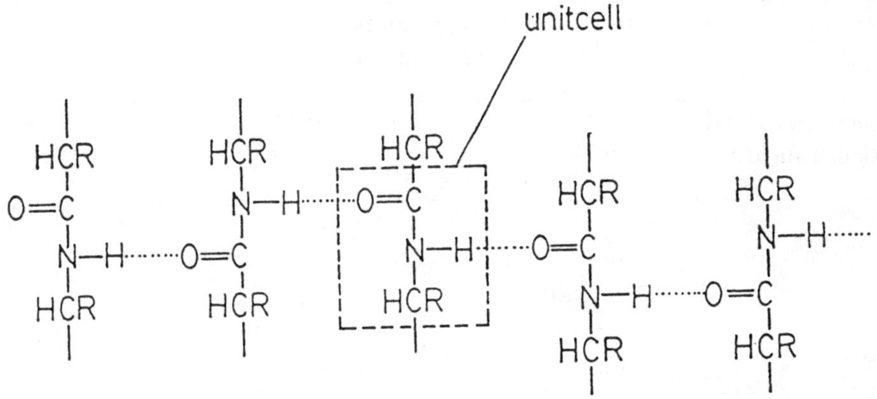

Fig. 5.14: The two-dimensional polypeptide network. The *dotted lines* again indicate hydrogen bonds. As one can see, if the side chain is always the same (homopolypeptides, like polyglycine, polyalanine, polyserine, etc.) the (NH)-(C=O)-(CHR) unit is repeated in the main chain. On the other hand the -(NH)-(C=O) unit is always periodically repeated in the cross chain independently from the nature of the different side chains. The *broken lines* surround a unit cell in the two-dimensional polypeptide network

The band widths shown in Table 5.2 correspond to the fact that the units in the main chain are much more strongly coupled (widths of 1.38 and 2.10 eV, respectively) than those interacting only through hydrogen bonds, which have widths smaller by about a factor of ten. The gap for the main chain is about 12.4 eV which is, again due to the factors discussed for the nucleotide base stacks, artificially far too large. The realistic value should be around 8 eV.

As the next step, the band structure of the main chain of polyalanine was calculated [2]. The widths of the conduction and valence bands, respectively, are somewhat smaller, i.e 1.2 instead of 1.4 eV and 1.8 instead of 2.1 eV, respectively. The upper limits of both bands are shifted downwards by about 0.25 eV. Figure 5.15 shows the positions of the valence and conduction bands of polyglycine and polyalanine, respectively.

Table 5.2: The widths of the valence and conduction bands of polyglycine (in eV-s) [2].

	main chain	hydrogen bonded perpendicular chain
Conduction Band	1.38	0.14
Valence Band	2.10	0.29

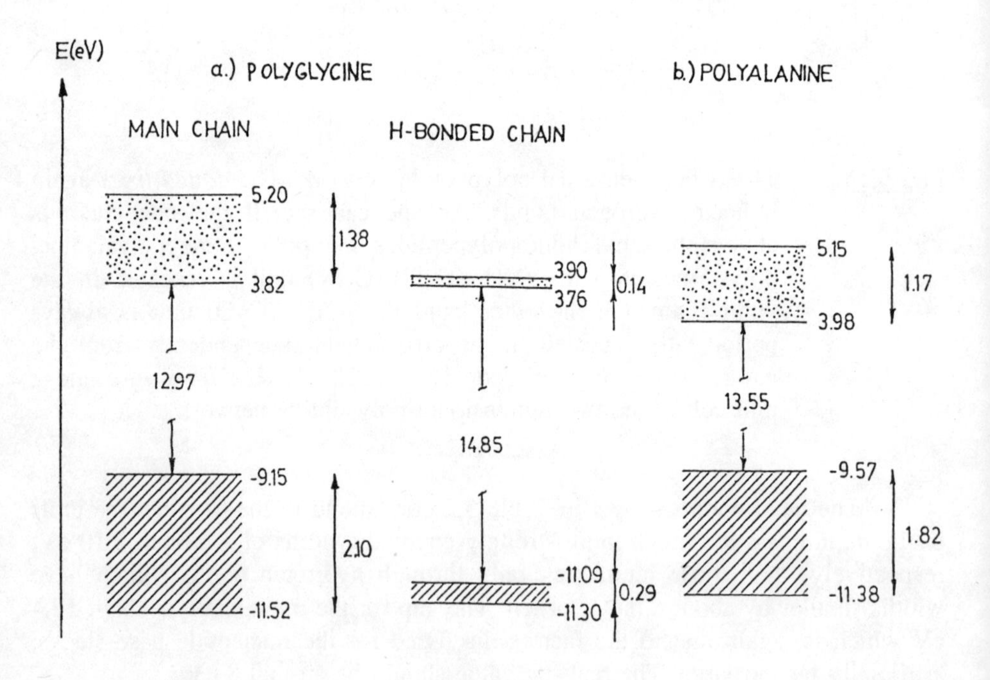

Fig. 5.15: The positions of the valence and conduction bands of polyglycine and polyalanine

Despite the rather similar chemical structures (alanine differs from glycine only by a substitution of a H-atom by a CH_3-group) and energy band structures, a calculation of an alternating glycine-alanine chain [poly(glyala)][26] resulted in the splitting of both the valence and conduction bands with gaps of 0.2 and 0.1 eV, respectively. Similar results have been obtained from the calculation of mixed glycine, alanine and serine chains [26,27] indicating the rather crucial role of disorder on the energy level distribution of proteins. At the moment, the calculation of the energy level distribution of an 8-component mixed polypeptide chain is in progress using rather advanced techniques. The 8 different components are chosen in a way that they should give a good representation of all the 20 amino acids. One hopes to obtain in this way a rather realistic picture of the level distribution of a real protein.

In polypeptide chains having 20 instead of 4 different components the disorder is certainly larger than in the nucleotide base stacks. On the other hand, due to the normal chemical bonding in the main backbone of a protein, the widths of the physically interesting valence and conduction bands of the homopolypeptides is substantially larger (by a factor between 3 and 8) than in the case of the periodic base stacks. This means that, despite the larger disorder, there still remain more or less continuous allowed-energy regions. The two-dimensionality of the problem (which has not yet been investigated in a disordered polypeptide) also facilitates the possibilities of higher electric conduction. To decide, whether a certain protein (with a given sequence of amino acids) can be only a rather poor or a better conductor, assuming again the presence of free charge carriers which can ge generated through charge transfer between a polypeptide chain and the sugar-phosphate chain, is in the same way a non-trivial problem as in the case of DNA (see, however, the remark on p. 125).

If we assume that carcinogens bind directly to proteins, or to the sugar-phosphate part of DNA in the close vicinity of the polypeptide chain in a nucleohistone, they will certainly increase further the disorder in the protein molecules and thus, via the same effects described in the discussion of DNA, will decrease the electronic conduction in them.

5.3.2 Different Long-range Effects of Carcinogens Bound to DNA or Proteins.

To understand the different long-range effects of carcinogens bound to DNA or proteins one has to keep in mind that the biological macromolecules already due to their long dimensions are complicated solids (see Sect. 5.3.1) and therefore the different long-range effects are essentially solid state physical effects.

One can show that the attractive dispersion interactions between two linear chains are orders of magnitude larger, if both are conductors and thus have partially filled valence bands, than if they are insulators, i.e. they possess completely filled valence bands and the fundamental gap is large [28]. This situation is shown in Fig. 5.16.

The dispersion interaction is attractive, because two neutral molecules or chains can mutually polarize the distribution of their electrons in a way that their electrostatic interactions becomes more negative (more attractive).

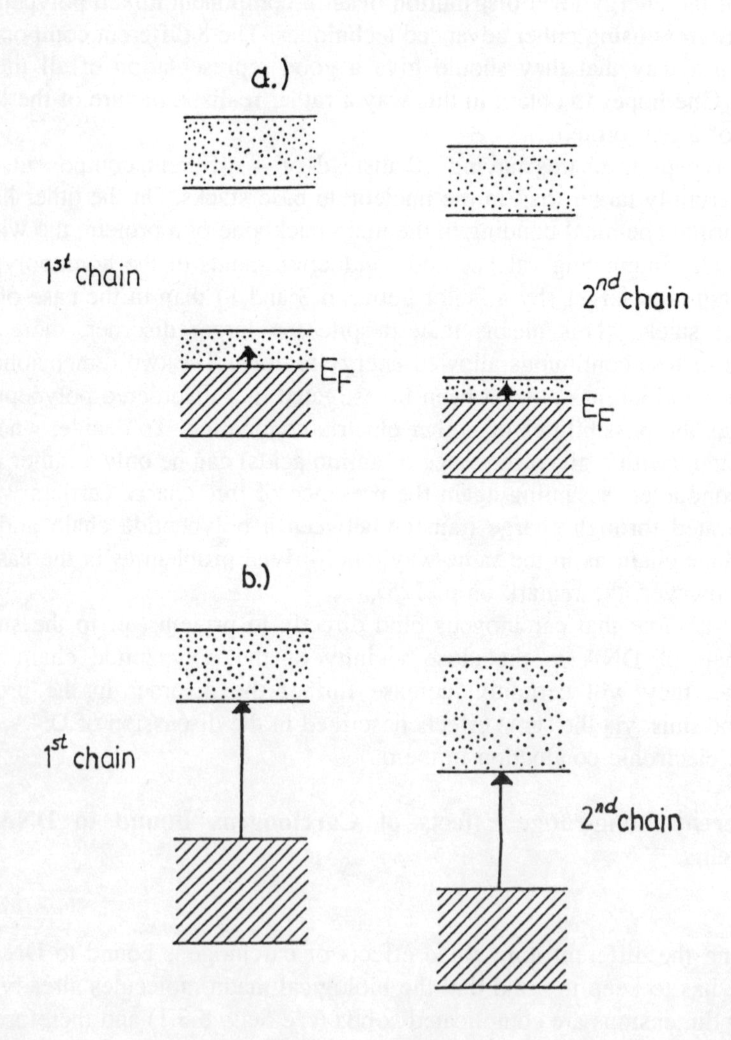

Fig. 5.16: The attraction between two conducting chains (a) is much larger than between two insulator chains (b). The *arrows* indicate possible excitations

In a rather good approximation of this phenomenon one has a ratio, in which the nominator is a constant for two given molecules in a fixed relative orientation, and the denominator contains excitation energies of the electrons of both molecules from their filled to their unfilled levels. If we make a linear chain from the molecules by binding them together, their individual energy levels as we have seen in the previous point broaden to energy bands. The excitation energies between these bands is somewhat smaller than between the corresponding energy levels due to the widths of the bands (see Fig. 5.16, part b), but until the bands are completely filled or empty, the dispersion energy between the two chains, calculated for a pair of unit cells, among which one belongs to one chain and the other one to the other chain will change only a little.

There is a completely different situation if both chains have only partially filled bands (see part a) of Fig. 5.16). In this case one can excite the electrons of both chains inside their valence bands with very little energy. This means that the denominator of the expression for the dispersion energy becomes very small and therefore the dispersion energy very large [28]. A detailed mathematical analysis of the problem which has taken into account also the difficulty that at the Fermi levels the denominator of the dispersion energy expression goes to zero, has shown that two conducting chains attract each other much stronger than two insulator chains [29].

It is rather probable that due to the internal charge transfer within the DNA molecule [13] and between DNA and protein [30], respectively, both the sugar-phosphate chain of DNA and the polypeptide chain of the protein surrounding it has a partially filled valence band (in the DNA case) and a partially filled level system in the conduction band region of the polypeptide. In consequence of the non-periodicity of the amino acids in a protein it has no continuous energy bands, but a dense distribution of energy levels in the different band regions of the periodic homopolypeptides (see point 5.3.1.). Let us assume now that a carcinogen bound to DNA or proteins causes a charge transfer and in this way either fills the valence band of the sugar-phosphate chain of DNA (electron donor carcinogens) or takes out the extra electrons from the conduction band region of the nucleoprotein (electron acceptor carcinogens). In both cases the denominator of the dispersion energy expression would considerably increase and therefore the attraction between the two chains due to the dispersion forces would decrease considerably.

Besides possibly having charge transfer effects carcinogens bound to DNA or nucleoproteins certainly increase the degree of disorder (non-periodicity) as mentioned already above in these chains. Though the sugar-phosphate chain of DNA is periodic, its different units see a non-periodic electric field due to the different nucleotide bases and to the different amino acids in the surrounding polypeptide chain (see Fig. 5.17). The disorder introduced in this way is, however, not very large because both the nucleotide base stacks and the polypeptide chain are displaced from the sugar-phosphate chain. On the other hand if a bulky carcinogen binds to a nucleotide base or to a sugar-phosphate unit directly, it causes a strong distortion of the local conformation. If a large number of carcinogens is doing this, this introduces an additional (and much higher) degree of aperiodicity than the previously mentioned ones. There is a similar situation in the cases if a methyl (CH_3-) or ethyl (CH_3-CH_2-) group is bound to the nucleotide bases. As it was mentioned in point 5.2 their binding introduces a net positive elementary charge at their sites of attack.

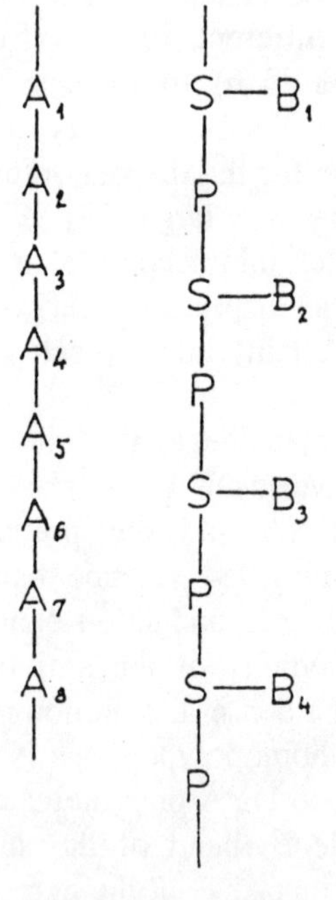

Fig. 5.17: Non-periodicity introduced in a periodic sugar-phosphate chain of DNA due to the neighbourhood of different nucleotide bases and amino acids in the polypeptide chain, respectively (schematic). The B_i-s stand for the different nucleotide bases, the A_i-s for the different amino acid units.

Both types of effects can cause such a degree of disorder that the band structure of the sugar-phosphate chain will be destroyed and that for instance the valence band decays into many very narrow bands or bunches of levels, respectively. This means that usually the denominator of the dispersion energy increases again and with it the attraction between the two chains again becomes weaker. In other words, increased disorder caused by carcinogens influences the DNA-protein interaction, making it less strong [2].

A third probable long-range effect on DNA-protein binding is shown in Fig. 5.18. In this case, which is easy to visualize, the folding of a DNA double helix (its tertiary structure) is changed because the carcinogen binding happens in such a way that the double helix does not fold back to the neighbourhood of the critical protein interacting with DNA. Therefore the interaction of the protein with DNA will have changed. The back-folding of the double helix has not only a direct effect on this interaction, but also the structure of water and the distribution of K^+ ions around the protein interacting with a gene of DNA will be different if there is back-folding or not. In this way, a carcinogen bound far away from the critical part of DNA (which interacts with a regulator protein) through the change of the tertiary structure has a long-range effect on the DNA-protein interaction under consideration [2].

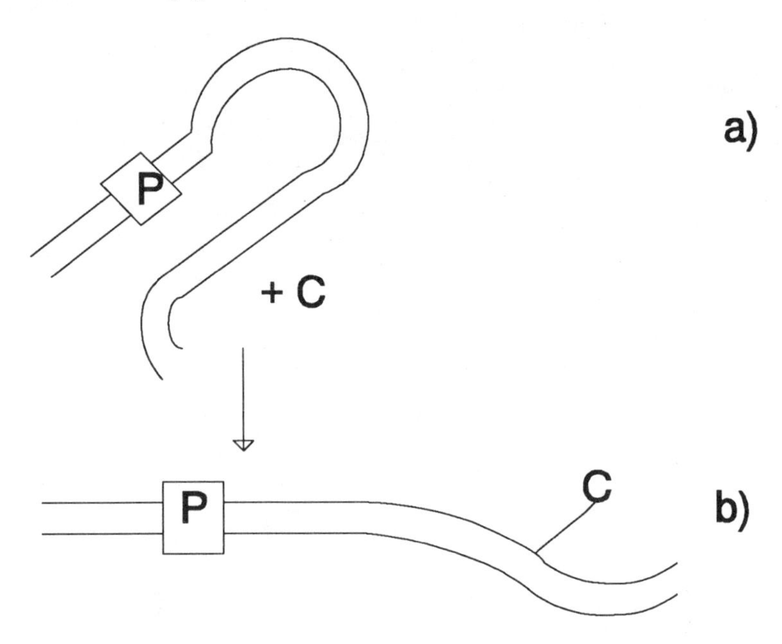

Fig. 5.18: A carcinogen bound to DNA (or protein) influences its folding (tertiary structure) and therefore the DNA-protein interaction changes, a) tertiary structure of DNA before carcinogen binding, b) tertiary structure of DNA after carcinogen (denoted by *C*) binding

As the last rather probable way carcinogens binding to DNA (or to protein) can influence the DNA-protein interaction through a long-range effect one should mention the formation of a so-called solitary wave (soliton). To understand this we show in Fig. 5.19 that if a bulky carcinogen binds to one of the nucleotide bases, this will locally strongly distort the geometrical arrangement of the stacked bases. For this stacking interaction which gives rise to the band structure of periodic base stacks see the Sect. 5.3.1. This means that the carcinogen causes, at its site of binding, a non-linear change, namely a geometrical change coupled with the change of the interactions between the electrons.

Until the carcinogen is bound to a DNA base this non-linear change will remain in the neighbourhood of its binding. Though carcinogen-nucleotide base bindings can be quite strong, under the conditions of a living cell (in vivo) the carcinogen can be cut off easily from the base, for instance by collisions with other molecules. When the carcinogen leaves the DNA double helix (see Fig. 5.20), it is rather improbable that the double helix will regain its original

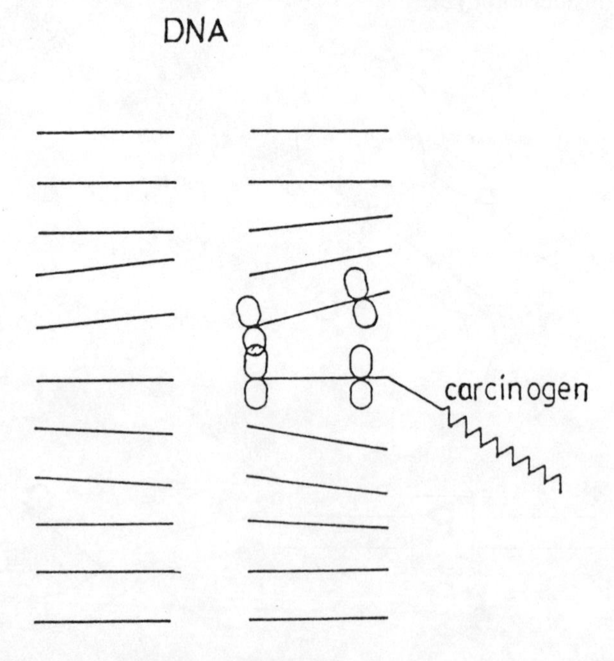

Fig. 5.19: The binding of a bulky carcinogen to a base in a nucleotide base stack causes (1) a geometrical distortion of the stack and (2) in consequence of this distortion a change in the interaction between the electrons belonging to different stacked bases (indicated by *overlapping ellipsoids*)

geometrical form (conformation) instantenously, because bulky molecules (the nucleotide bases) together with their surrounding water molecules and ions would have to be moved instantenously for this purpose. It is much more probable that the above described non-linear change starts to travel along the chain in both directions as indicated by arrows in Fig. 5.20 [31]. Recent rather complicated quantum theoretical calculations have shown that this really is the case [32].

Non-linear solitary waves corresponding to an excited state of the system, have two basic properties: (1) as compared to simple electronic excitations they have much a longer lifetime, i.e. the time before the system returns into its ground state, while solitons in a strict sense would have an infinite lifetime and (2) they can travel along a chain [33]. In the case of a nucleotide base stack a solitary wave can most probably travel along the stack for a larger distance influencing the DNA-protein interactions on its way. To have, however, a lasting effect on these interactions, it has to give away energy to be able to break

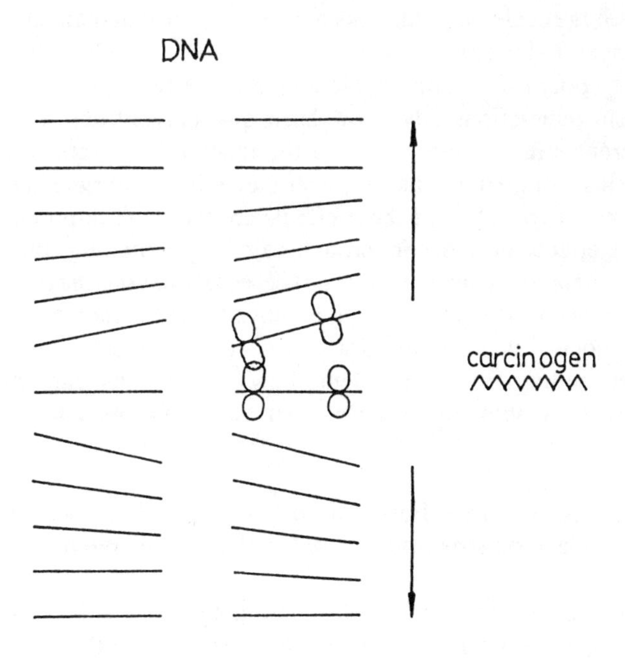

Fig. 5.20: After the cut off of the carcinogen the coupled geometrical distortion and electronic interaction change starts to travel along the chain (indicated by *arrows*) as a non-linear so-called solitary wave

hydrogen bonds between DNA and proteins or/and weaken other electric interactions between these two kinds of chains. One would think therefore that a solitary wave will die out after travelling along the stack after a few base pairs. This would be true if DNA were in vacuum. If we consider, however, the possibility of the rearrangement of the water molecules and ions around DNA (which always surround it in a living cell) the situation is completely different.The readjustments of the water structure and ion distribution accompanying the solitary wave can refuel it with energy and so the energy which is given away by it by weakening the DNA-protein interactions, can be regained from the very large energy reservoir formed by the water and ion environment of DNA. Therefore one expects that the described solitary wave is able to travel along a nucleotide base stack and can weaken its interactions with proteins at a longer segment of DNA. To answer the question whether this is really so, an appropriate mathematical description which also takes into account the time-dependence of the phenomenon described has been developed [32] and on its basis large scale computations have been performed and are still in progress in the Erlangen Laboratory.

Four different probable long-range effects of carcinogen binding on DNA-protein interactions have now been qualitatively described. Three of these four different effects were based on the quantum theoretical treatment of DNA and proteins using solid state physical methods. To answer the fundamental question which one (or ones) of these effects are the most important or which other long-range effects may occur through carcinogen binding, one has to continue large scale theoretical investigations. These calculations have to be supplemented, of course, with the corresponding biophysical experiments (see in detail in Sect. 8.6). One hopes, that in this way, one would obtain a much more fundamental understanding of the problem, how carcinogens can influence DNA and its interactions with proteins (in other words the genetic regulation in eukaryotic cells).

5.3.3 Connections Between the Different Long Range Physical Effects of Carcinogens and the Activation of Oncogenes.

In the preceding section, different long-range solid state physical effects of carcinogens bound to DNA and/or proteins were described. It was shown in all cases that these effects influence (usually weaken) the DNA-protein interactions. One should not forget that the interaction between two molecular chains, like DNA and proteins, consists of quite a few different attractive and repulsive interactions. If one such giant molecule is bound to another one, then the sum of the attraction energy terms is larger than the sum of the repulsion ones. In many cases the difference between these two sums is quite small as compared to the values of the individual terms. This means that there is a delicate energy balance in favor of the binding (attraction) and a comparatively small decrease in the attraction terms can shift this energy balance into the other direction: the sum of the repulsion terms will be the larger one. If this happens instead of binding the two molecular chains will separate from each other.

On the basis of the above sketched considerations, it can be easily understood that through long-range carcinogen effects the net DNA-protein interaction will become repulsive at an oncogene or at the DNA segment which regulates the use of the genetic information of the oncogene for protein synthesis. The oncogene which has become in this way deblocked from its protein "coat" can e.g. be more easily approached by those enzymes which are necessary for slicing out the whole oncogene from a less active chromosome and build it into another more active one, i.e. in other words the second chromosome is regulated so that the genetic information contained in the oncogene is more frequently expressed. As described in Chapter 2, this results in an overactivation of the otherwise normal genes, so that the corresponding proteins are produced in much larger quantities, which can amount up to a factor of 1000 of the quantity found in normal cells. The same holds for the amplification and LTR ligation mechanisms also described in Chapter 2. This overproduction of (presumably) regulatory proteins can destroy then the self-regulation of the cell and in this way start its cancerous change (see Chapters 2 and 6).

5.4 The Effects of Radiation on DNA

It is well known that different forms of radiation such as UV and ionizing radiation, i.e. X-ray or particle radiation can initiate cancer. These forms of radiation act mostly indirectly by producing first of all alkyl radicals (see p. 61) which then attack DNA by binding to its different constituents. Therefore in this case, the indirect effects of radiation can be understood on the basis of the same mechanisms which were discussed in Sects. 5.2 and 5.3 for chemical carcinogens.

At this point, we are going to discuss the different effects of direct-hit radiation on DNA. The primary mechanism of direct hit UV and particle radiation on DNA have unfortunately become very important in the last decade. First, there is the increasing size of holes in the ozone layers around the extended polar regions of the Earth, including recently also Canada and the New England part of the United States in the northern hemisphere. Secondly, there is the dangerous possibility that with the proliferation of atomic reactors, especially in some countries without sufficient safety measures, catastrophes like that of Chernobyl may occur again.

In this situation, it is mandatory to understand as much as possible the effects of these radiations on the genetic material on the basic electronic level. Only in this way, can we hope to understand the final biophysical phenomena which are due to radiation. Such understanding can most probably provide us with the combined chemical, biological and physical tools to prevent and counteract the final lethal effects of higher doses of UV or ionizing, high energy radiation.

5.4.1 Direct Hit UV Radiation

UV direct hits on DNA usually have three primary effects, depending on the energy of the photons: excitations, ionizations and pyrimidine-pyrimidine (T or C) photodimerizations (see pp. 61 and 111). The $\pi \rightarrow \pi^*$ excitation energies of the single nucleotide bases have been found experimentally to be around 4.5 eV. Theoretical calculations starting from the band structures of the nucleotide base stacks (see Sect. 5.3.1) are in reasonable agreement with the experimental results. E.g. for a cytosine (C) stack the first excitation occurs according to the calculations at 4.7 eV [34]. Both experiment and theory indicate that the primary excitation occurs rather locally. Ionization and T-T or C-C dimerization are still more localized primary events than excitation. Through several chemical reaction steps, such a pyrimidine-type nucleotide base dimerization can easily lead finally to the break of the DNA strand on which the dimerization occurred [35].

In all these cases, a geometrical (conformational) change occurs which causes also a local change of the electronic structure especially due to the changed electronic stacking interactions. In the case of the pyrimidine-pyrimidine photodimer the conformational change is obvious. An excited molecule, as is well known, has a geometry different from that in its ground state. In the case of benzene and some other ring molecules it has been experimentally established [36], that in their excited states, the "radius" of these ring shaped molecules increases. Though, according to the authors' knowledge, no such measurements have been performed on the nucleotide bases, it is justifiable to assume a similar increase of the bond distances in these ring molecules (see Fig. 5.21). However, in the case of ionization, owing to the net positive charge one would expect an opposite geometrical change in the nucleotide bases: the positions of their atoms will be shifted towards the "middle points" of the rings. To prove these geometrical changes, calculations are in progress on a C stack.

In the three different above-mentioned cases, we have a local non-linear effect: a conformational change coupled with a change in the local electronic structure inside the stack. Of course the conformational changes described also induce local changes in the water structure and ion distribution around DNA.

Let us now assume that the primary effect of the UV photon hit relaxes. This means that the photo dimer breaks off, which in vivo can occur much more easily than <u>in vitro</u>, the ionized nucleotide base recaptures an electron, or the locally excited state of the base stack returns to its electronic ground state in a radiationless way by distributing its excitation energy among different vibrations.

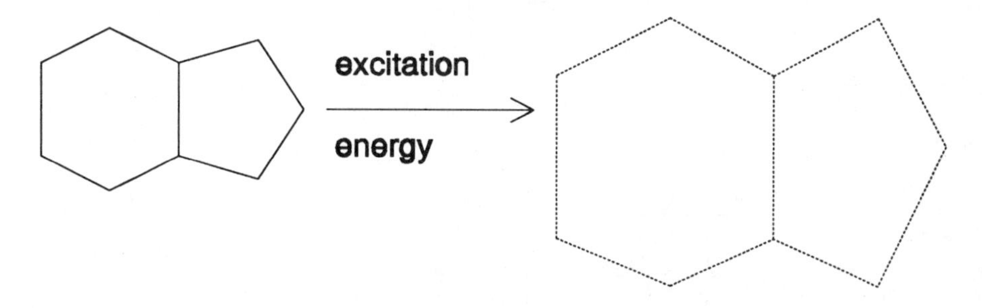

<u>Fig. 5.21</u>: A ring shape molecule (on the left hand side in its ground state) will become larger in its excited state (right hand side)

The question arises again whether the above described local non-linear change also relaxes instantaneously, or whether moving solitary waves, i.e. coupled conformational and electronic structure changes together with the changes of the environment of DNA, which can have a very long life time and move along the stack, are generated.

Our calculations on a nucleotide base stack as mentioned before [32] have shown that the latter is the case, if one changes the position of a base within the stack and keeping the geometries of the individual molecules unchanged. One should point out that although the four nucleotide bases are rather different, the two different base pairs are more similar because each one has a purine and a pyrimidine type ring and the number of π electrons in both pairs is the same. Therefore one can expect that a solitary wave can travel along even a non-periodic double stranded DNA. Non-periodicity of the stacked base pairs causes energy losses of course, but the energy of the solitary wave can be replenished by the cytoplasm surrounding a nucleoprotein acting as a heat bath.

To demonstrate the occurrence of solitary waves caused by UV radiation, calculations are in progress involving an appropriate change in geometry of one nucleotide base in a DNA B stack.

One should mention further that an excited molecule in a stack does not return necessarily to its electronic ground state: its excitation energy can hop between the different molecules in the stack. In this case again, a non-linear distortion (here a geometry change in the molecule coupled to an electronic structure change due to the excitation, to the difference in stacking interactions and to accompanying environmental changes) is travelling as a solitary wave along the nucleotide base stack (see Fig. 5.22). The corresponding calculation for a DNA base stack is also in progress in Erlangen.

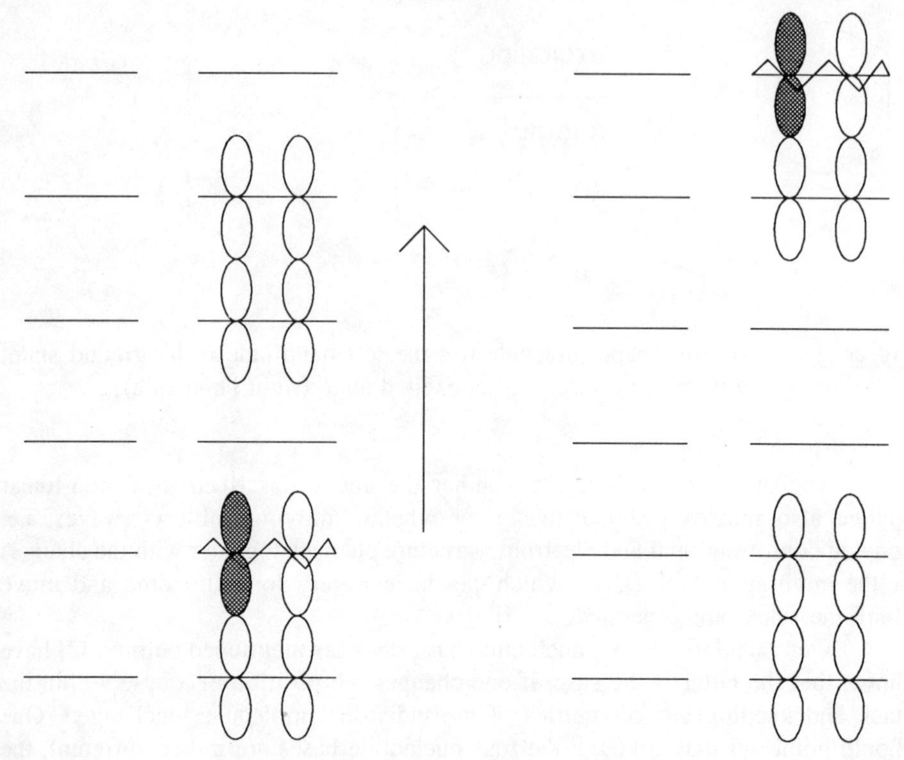

Fig. 5.22: Formation of a solitary wave in a stack because of the simultaneous change of the size of the excited molecule (indicated by a *zig-zag line*) and the change of the electronic interaction of the excited molecule and its neighbour (indicated schematically by *shaded orbital lobes* instead of *non-shaded ones*)

Finally, a word has to be said about the effect of electromagnetic radiation with much higher energy than UV radiation (X-ray or γ radiation). In this case, besides the physical mechanisms of oncogene activation sketched above, single and double strand breaks of DNA play a very important role (see Sect. 5.4.3). If a double strand break occurs in DNA, the loss of genetic information again can have a carcinogenic effect by the loss of the regulatory genes, the so-called antioncogenes, which suppress excessive DNA duplication.

5.4.2 Particle Radiation

If DNA suffers direct hits from charged particles, electron energy loss experiments have shown that probably the most important primary event apart from strand breaking(s) is the collective excitation (plasma excitation) of the whole valence electron system of the base stacks, with an energy of about 21 eV. In these experiments DNA is bombarded by high energy (25 KeV) electrons, and the energy lost on average by the individual electrons because of their interaction with DNA is measured. One finds in the energy loss spectrum a smaller peak at about 6 eV and a more intensive broader one at about 21 eV in all the five bases [37,38]. In other aromatic organic solids, the spectrum is very similar to that of the nucleotide bases [39].

One should mention that the collective (plasma) excitation of an (in principle) infinite electronic system means that each electron of the system is excited by an (in principle) infinitely small amount of energy. One can visualize this by assuming a large but finite number of electrons N and a collective excitation of the whole electronic system by a finite amount of energy (E_{coll}). Then the excitation energy of a single electron is equal to E_{coll}/N. If $N \rightarrow \infty$ (which in practice is never the case, only N is large), this ratio would tend to zero. It should be pointed out that according to the theory of collective excitations of large electronic systems E_{coll} is not a function of the number of electrons N.

In a pilot calculation in 1968 we calculated (using a simple theory of Pines [40]) the peak to be around 6 eV [41] (which is interpreted as the collective excitation of the π electron system [41] of the nucleotide bases) applying the π electron band structures of different periodic nucleotide base and base pair stacks available at that time [42]. These calculations provided values between 5.7 and 6.1 eV for the energy of the first collective excited state (depending on which stack had been investigated), in quite good agreement with experiment.

Band structure calculations to compute the second collective excited state (all-valence electron excitations) of a C stack have been performed. In this case the same method was used as that applied [15] to obtain the quantities given in Table 5.1 on p. 116 and the band structures obtained in this way were corrected for the finer details of the interelectron interactions [43]. Further to obtaining the energies of the collective excitations, the above-mentioned simple theory of Pines [40] has been applied again, but in this case for all the valence electrons. The results obtained [44] showed a very intensive peak at about 19 eV for a C stack and a shoulder at 17 eV in good agreement with the experimental values of 21 and 16 eV [37], respectively.

It is known that a collective excited state of a solid or a polymer (like DNA) decays with a very short lifetime into a number of single excitations and ionizations. The detailed quantum mechanical theory of these transitions has not yet been worked out completely, though some work has been done in this direction [45].

After the decay of the collective excitations into single particle excitations (excitons) and ionizations in the base pair region of DNA, one can apply the same arguments as in the case of UV radiation to show that again at different points of the base stacks solitary waves are generated. These solitary waves can again contribute both to the long range effects necessary for oncogene activation and to double strand breakings in DNA which can inactivate antioncogenes.

5.4.3 Qualitative Theory of Double Strand Breaks and Inactivation of Antioncogenes

If a high energy radiation quantum (X-ray photon or particle) hits DNA first, most probably a single strand break occurs in DNA besides the collective excitation described above in the case a charged particle hits (see Fig. 5.23).

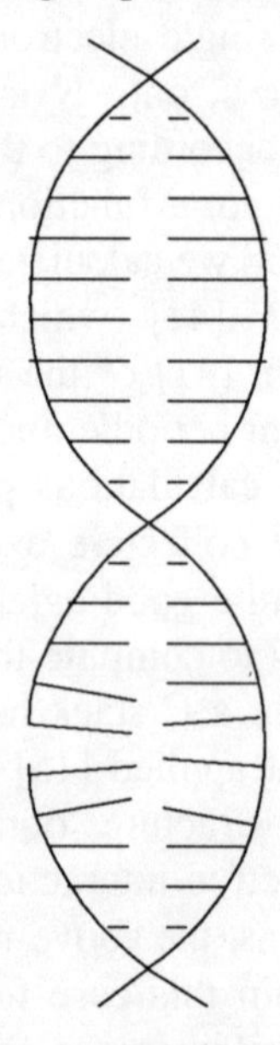

Fig. 5.23: Single strand break caused by a single hit of radiation. Notice that the position of the nucleotide bases is disturbed also in the unbroken second strand

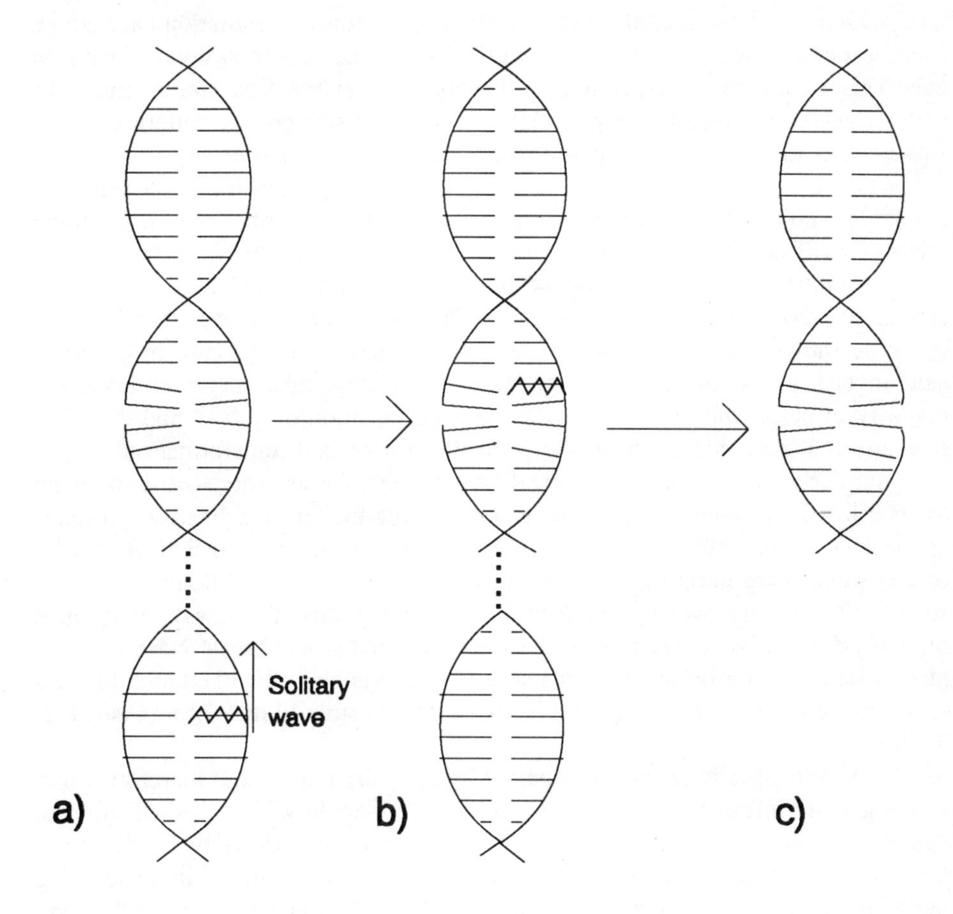

Fig. 5.24:	a)	A nucleotide base becomes excited far away from a single strand break (indicated by a *zig-zag line*)
	b)	The solitary wave caused by the excitation of a base arrives in the other strand at the place, where the first chain was broken
	c)	The solitary wave has lost its energy which has caused a break in the second chain. A double strand break occurs in this way

One could assume that if the quantum of the ionizing radiation has enough energy, it could break the second strand of DNA in the neighbourhood of the first one. This is, however, rather improbable, because at the first strand break the particle gets scattered and the probability of forward scattering (scattering with a small angle relative to its original direction) is, as one knows from the theory of scattering, small. This is reflected also by the first equation on p. 66 (see also references [86-89] of Chapter 3) according to which the number of double strand breaks (DBS) denoted there by Y depends quadratically on the dose D.

Another possibility would be that two subsequent particles hit the two strands of DNA very close to each other. This event is again very improbable as it can be shown with similar statistical considerations as in the case of chemical carcinogens, unless the dose is very large. In the case of a very high dose, the majority of cells will be destroyed (see second equation on p. 66) and therefore this case is not of interest from the point of view of cell transformation. In this way, similarly to the case of chemical carcinogens, we are forced to look again for possible long-range mechanisms to explain the higher than linear occurrence of DSB-s (Y) as a function of the dose (D). On the other hand if a second hit occurs somewhere quite far away from the site of the first one along the second strand, the solitary wave generated in the above described way, will most probably deposit its energy near to the site of the first strand break. Namely at the place where the conformation of the second strand is also perturbed and this acts as a trap for the solitary wave. In this way a double strand break occurs (see Fig. 5.24).

A DSB usually cannot be repaired by a repair enzyme and therefore a part of the genetic information carried by the DNA molecule will be lost. If this lost genetic information contains antioncogenes, they cannot exert their regulatory function through the proteins coded by them. In this way, direct hits of ionizing radiation coupled with long range effects along DNA can again cause a strong disturbance in the self-regulation of a living cell. This can lead then again to the initiation of a malignant transformation of the cell.

References

1. H. Busch, in Introduction to the Biochemistry of the Cancer Cell, Academic Press, New York, London, 1962.
2. J. Ladik, S. Suhai and M. Seel, Int. J. Quant. Chem. **QBS5**, 35 (1978).
3. F.A. Cotton, V.W. Day, E.E. Hazen, Jr. and S. Larsen, J. Am. Chem. Soc. **95**, 4834 (1973).
4. R.S. Day, F. Martino and J. Ladik, J. Theor. Biol. **84**, 651 (1980).
5. D. Grunberger, personal communication, 1992.
6. J. Ladik, Physiol. Chem., Phys. and Med. NMR, **22**, 229 (1990).

7. I.B. Weinstein, A.M. Jeffrey, K.W. Jenette, S.H. Blobstein, R.G. Harvey, C. Harris, H. Autrup, H. Kasai and K. Nakanishi, Science **193**, 592 (1976).
8. I.B. Weinstein and D. Grunberger, Chemical Carcinogenesis, P. Tso and J. Dipaolo eds., Marcel Dekker Inc., New York, 1974, Part A, p. 217.
9. See for instance: E. Santos, S.T. Tronick, S.A. Aaronson, S. Pulciani and M. Barbacid, Nature **298**, 343 (1982). Further references see in Chapter II.
10. R.B. Setlov, Progress Nucl. Acid Research, Mol. Biol. **8**, 257 (1968).
11. C. Kittel, Introduction to Solid State Physics, J. Wiley, New York, 1971.
12. J. Ladik and S. Suhai in Theoretical Chemistry, ed. C. Thomson (Royal Soc. of Chemistry), Vol. 4. 1981, p. 49; J. Ladik, "Quantum Theory of Polymers as Solids", Plenum Press, New York, London, 1988.
13. J. Ladik and S. Suhai, Int. J. Quant. Chem. **QBS7**, 181 (1980).
14. J.F. Gentleman, M.A. Shadbolt-Forbes, J.W. Hawkins, J. Ladik and W. Forbes, Mathematical Scientist **9**, 125 (1984).
15. P.-O. Löwdin, Adv. Phys. **5**, 1 (1956); G. Del Re, J. Ladik and G. Biczó, Phys. Rev. **155**, 992 (1967); J.-M. André, L. Gouverneur and G. Leroy, Int. J. Quantum Chem. **1**, 427, 451 (1967).
16. C.C.J. Roothaan, Rev. Mod. Phys. **23**, 69 (1951).
17. S. Suhai, Phys. Rev. **B27**, 3506 (1983).
18. S. Suhai, Int. J. Quant. Chem. **QBS11**, (1985).
19. R.S. Day and J. Ladik, Int. J. Quant. Chem. **21**, 917 (1982).
20. P.W. Anderson, Phys. Rev. **109**, 1492 (1958).
21. Y. Ye and J. Ladik, Phys. Rev. **B48**, 5120 (1993), and references therein.
22. N.F. Mott and E.A. Davies, Theory of Non-Crystalline Solids, Clarendon Press, Oxford, 1971, p. 215.
23. E. Clementi, Computational Aspects for Large Chemical Systems, Lecture Notes in Chemistry, Springer, London-New York-Heidelberg, 1980, Vol. 19; E. Clementi, G. Corongiu, M. Gratanola, P. Habitz, C. Lupo, P. Otto and D. Vercauteren, Int. J. Quant. Chem. **S16**, 409 (1982).
24. P. Otto, J. Ladik, G. Corongiu, S. Suhai and W. Förner, J. Chem. Phys. **77**, 5026 (1982).
25. S. Suhai and J. Ladik, Acta Chim. Sci. Hung. **82**, 67 (1974).
26. S. Suhai, J. Kaspar and J. Ladik, Int. J. Quant. Chem. **17**, 995 (1980).
27. R.S. Day, S. Suhai and J. Ladik, Chem. Phys. **62**, 765 (1981).
28. K. Laki and J. Ladik, Int. J. Quant. Chem. **QBS3**, 51 (1976).
29. F. Beleznay, S. Suhai and J. Ladik, Int. J. Quant. Chem. **20**, 683 (1981).
30. J. Ladik, Int. J. Quant. Chem. **QBS3**, 237 (1976).
31. J. Ladik and J. Čížek, Int. J. Quant. Chem. **26**, 955 (1984).
32. D. Hofmann, J. Ladik, W. Förner and P. Otto, J. Phys. Condens. Matter **4**, 3883 (1992); J. Ladik, D. Hofmann, W. Förner and P. Otto, Physiol. Chem., Phys. and Med. NMR, **24**, 227 (1992).

33. See for instance: A.S. Davydov and N.J. Kislukha, Phys. Status Solidi **59**, 463 (1973); A.S. Davydov, Phys. Scripta **20**, 387 (1979).
34. S. Suhai, Int. J. Quant. Chem. **11**, 223 (1984).
35. Osman, Int. J. Quant. Chem. **QBS14** (in press).
36. J.N. Murrell, The Theory of the Electronic Spectra of Organic Molecules, Methuen and Co. Ltd, London, 1963, point 6.3; C. Longuet-Higgins and L. Salem, Proc. Roy. Soc. **A251**, 172 (1959).
37. C.D. Johnson and T.B Rymer, Nature **213**, 1045 (1987).
38. M. Isaacson, J. Chem. Phys. **56**, 184 (1972); C.D. Johnson, Radiation Research **56**, 63 (1972).
39. N. Swansson and C.I. Powell, J. Chem. Phys. **39**, 630 (1963).
40. D. Pines, Rev. Mod. Phys. **28**, 184 (1956).
41. J. Jäger and J. Ladik, Phys. Lett. **28A**, 328 (1968).
42. J. Ladik and K. Appel, J. Chem. Phys. **40**, 2470 (1964); J. Ladik and G. Biczó, J. Chem. Phys. **42**, 1658 (1965).
43. J. Ladik, A. Sutjianto and P. Otto, J. Mol. Struct. (Theochem) **228**, 271 (1991).
44. J. Ladik , H. Früchtl and P. Otto, J. Mol. Struct. **297**, 215 (1993).
45. I. Egri, Excitons and Plasmons in Metals, Semiconductors and Insulators: a Unified Approach, Phys. Reports, **119**, 363 (1985).

6 The Disturbance of Cell-Self-Regulation Due to Oncogene Activation and Antioncogene Inactivation

6.1 Changes in Protein Structures and Concentrations Due to Oncogene Activation and Antioncogene Inactivation

We mentioned in Chapter 2 that point mutations in a protooncogene can cause a change in the sequence and possibly also in the conformation of the protein coded by the oncogene. If this so-called oncoprotein takes part in the self-regulation of the cell, it can disturbed to such an extent that the cell goes over to another stationary state. Namely a living cell is not in thermodynamical equilibrium, because it is, due to the metabolism, in contact with its environment, and thus an open system. Therefore one can only speak about so-called stationary states of a cell in a given phase of its duplication cycle in which the quantity of the different chemical substances which enter and leave it per time-unit does not change. Further its temperature, volume, the pressure inside the cell, and the osmotic pressures of the transport of chemicals through its various membranes is constant. An open system can exist in different stationary states. If a living cell in its normal state is in a given stationary state and if instead of a protein taking part in its self-regulation a changed protein (an oncoprotein) occurs, or the original regulatory protein becomes inactive because of a point mutation, the cell can go over to another stationary state which is characterized by other flow of chemicals (its metabolism changes) and/or different thermodynamical parameters. The transition of a cell from its normal stationary state to another one can cause the initiation of the malignant transformation of the cell.

A classical example for the above described event is the case of the C-H-*ras* oncogene which was the first human oncogene discovered (see Chapter 1). If in the corresponding protooncogene the point mutation

$$\begin{matrix} G & & G \\ G & \rightarrow & T \\ C & & C \end{matrix}$$

occurs, in the protein coded by this gene the amino acid residue glycine is substituted by valine. The calculation of the spatial arrangement (conformation) of the mentioned protein with the help of empirical potential functions (whose parameters were determined experimentally) has shown that the single gly → val amino acid residue change brings the protein into a completely different conformation [1]. Therefore one can easily understand that the protein can no longer carry out its regulatory function, the biological, especially enzymatic activity of proteins depends on the sequence of amino acid residues as well as on its conformation. In this way the original self-regulation of the cell changes and the cell goes over to another stationary state which is the first step to its malignant transformation. The malignant transformation of a cell is a multistep process which usually needs the activation of at least two different oncogenes.

As was discussed in Chapter 2, the activation of oncogenes through chemicals happens in most cases not by point mutations, but by other biochemical mechanisms Such mechanisms are, e.g., insertion of long terminal repeat (LTR) segments, jumping of a regulatory gene or gene cluster from one chromosome to another, repetition of the same gene many times in the same strand of DNA, etc. (for details see Chapter 2). In all these cases, as discussed before, the primary sequence of a gene coding a regulatory protein does not change, but its regulation, i.e. its availability for protein coding, does so to a great extent. In most such cases, the protein coded by the gene is produced in an amount up to thousand times its original value. As was discussed in detail in Chapter 5, such an overactivation of a gene coding a regulatory protein can be understood only if one assumes long-range solid state physical effects which cause the release of the proteins blocking the gene. In this way the gene can code through RNA the corresponding protein for a much longer period span or many more times than in the normal state of the cell. It is easy to understand that if certain regulatory proteins are present in a 1000 times larger amount than usual and also at an unusual time point of the cell duplication cycles, this again causes such a strong disturbance of the self-regulation of the cell that it goes over from the normal to another stationary state (initiation of carcinogenesis).

In a similar way, as it was discussed also in Chapter 5, if different radiations cause double strand breaks of DNA, again via long-range solid state physical mechanisms, this causes a partial loss of the full genetic information. If in the lost part of the information carried by DNA some antioncogenes were situated, which are regulatory genes coding such proteins which hinder uncontrolled cell duplications, the cell can again go over to a state which differs from the normal stationary state. This can again initiate its malignant transformation.

As mentioned in Chapter 1, we know nowadays about 100 human oncogenes. However, in principle any regulatory gene, if it becomes overactivated can initiate the cancerous change of a cell. This overactivation can be caused by chemicals, or it is possible that the gene can no longer carry out its function, due to double strand breaks caused by radiation leading to a loss of parts of the genetic information which contain antioncogenes. In the human body there are about 5000 genes which have some regulatory function [2] which means that the known 100 human oncogenes may form only the tip of the iceberg.

6.2 Qualitative Description of a Mathematical Model for the Chemical Reactions Characterizing Living Cells and Their Ensembles

6.2.1 General Considerations

We have started to prove our assumption that the initiation of the malignant transformation of a cell starts by a change from its normal stationary state to another one, which is caused by overproduction of regulatory proteins or by their inactivation. These regulatory proteins can become inactivated either by losing their activity through conformational changes, or by their complete disappearance because of the loss of genetic information (antioncogenes) due to double strand breakings caused by radiation. For this purpose one has to study in more detail the regulation of cells and cell ensembles, than it was done in the past. To achieve this one has to investigate the reactions inside a cell, the interactions between cells, and their regulation of metabolism and cell growth. To this end, one should use all available mathematical theories, such as information theory, regulation theory (cybernetics, game theory, system theory etc.).

For an improved application of these theories one has to try to construct suitable mathematical models for the regulation of the metabolism of cells. These models have to contain the characteristic features of a living system (e.g. homeostasis) and have to take into consideration the essential non-linearity of biological systems. Furthermore, one should describe the most essential features of the biochemical reaction cycles in the cell using reaction kinetics. The velocity of a chemical reaction depends on a constant, the so-called rate constant, and on different powers of the concentrations of the molecules taking part in it. In this context the coupling of reaction cycles to structures with a higher organization, as in the theory of hypercycles introduced by Eigen and Schuster [3] for the description of self-replicative systems, seems to be very interesting.

The systems of differential equations which describe a biochemical reaction network are too complicated to be solved analytically, therefore, they have to be treated numerically. To characterize the dynamic behavior of the systems under investigation, a qualitative analysis of the systems of equations and their solutions can be performed with appropriate mathematical tools [4].

In building the mathematical model the presence of structures far from equilibrium in the sense of Prigogine and Glansdorff [5] has to be taken into account. The concept of self-organization of non-equilibrium systems [6], and the result that non-equilibrium can be a source of order are essential to describe living systems with the help of the laws of physics. In order to build the mathematical models one has to collect the most important data on matter, energy, and information transport within a cell and between cells. For understanding the regulation within a cell one should use the results of quantum mechanical investigations of energy and charge transport [7] in biopolymers and the results obtained from the interactions between biopolymers [8].

Another task is the development of a theory which describes how the behavior of a cell is regulated through interactions with all the other cells in an organism. For that purpose Haken's theory of cooperative phenomena in multicomponent systems (synergetics) [9] is applicable. One has to investigate the different kinds of possible matter and energy transports between cells through which the exchange of information and mutual regulation takes place. Such theoretical investigations should lead to regulation criteria which define the difference between cell ensembles of normal and tumor cells. A successful synergetic model of cell ensembles should finally lead to an interpretation of experimental tumor growth curves.

One has to keep in mind that regulation theory in its current form was essentially developed for the description of linear electric networks. For biological systems, on the other hand, the non-linearity of the kinetics is absolutely necessary in order to obtain a stationary state of an open system. Therefore, it is imperative to develop further regulation theory itself for the description of essentially non-linear processes. The solution of this purely mathematical problem would very probably yield important new insights into the regulation of metabolism and growth of cells.

6.2.2 Description of a Single Cell

As a first step towards the understanding of the regulation of a cell and of cell ensembles, the basic equations which describe a single cell can be written down. Further it can be outlined how the concept of cell compartments can be used to solve the arising system of coupled equations.

A single cell can be viewed as a highly organized conglomerate of different multienzyme systems. It is no longer regarded as a system composed of an aqueous solution in which enzymatic proteins and freely diffusing metabolites are uniformly dispersed. The modern concept of cellular metabolism is that enzymes operate within well-defined spatial regions [10]. The idea of a structural-functional organization helps for the segregation of competing reaction pathways, for the maintenance of a low average, but high local concentration of intermediate intracellular metabolites, and for the reduction of the transient time of a given chemical. Of course, the different regions (compartments) are coupled. Due to this coupling, mutual stabilization and regulation (inhibition or activation) can be affected by a single compound which acts only in a few neighboring compartments.

This spatial structure, i.e., different regions that are only weakly coupled, whereas inside a given region certain reactions are strongly coupled, can help to solve the rather complicated non-linear system of equations which should describe the whole cell. One can use here the following idea: It is assumed that a single cell can be described by a system of equations. These equations have to describe the kinetics, i.e. the velocity of the most important chemical reactions in the cell, the transport of matter (diffusion), the metabolism of the cell with its surroundings, the synthesis of DNA and RNA, oxygen metabolism, and protein synthesis, selfduplication, etc. The first problem is the choice of subsystems, like cell nucleus, mitochondria, ribosomes, Golgi bodies, cytoplasma, cell membranes, etc., their volumes, the reactions which take place in them, and the choice of interactions between these subsystems. In addition, appropriate rate and diffusion constants have to be assigned. First one could take model parameters for the reaction and transport processes. In the future these parameters should be calculated from microscopic models or obtained from measurements.

A cell cycle has four different phases [11]: mitosis (cell division), G_1 (pre-DNA synthesis), S (period of DNA replication), and G_2 (post-DNA synthesis). Each phase has to be described by a different system of equations.

Taking reasonably well-estimated reaction rate- and diffusion constants for each chemical reaction, one can write down the generalized kinetic equation [5] for each chemical reaction which one intends to include in the cell model. The selected chemical reactions can be divided according to the assumed spatial regions (compartments). In this way one solves the system of equations in each phase of the cell duplication cycle only for one compartment at a time and uses the results obtained for the concentrations of the reaction products as input data for those compartments which are coupled to the first one. Solving the equations individually for each compartment and repeating this procedure until the concentrations no longer change, one obtains so-called consistent solutions for the concentrations of all the different chemical species which take part in the selected chemical reactions.

In a real cell there are about 50,000 chemical reactions, but possibly one can build up a model cell with a few hundred reactions selected in such a way that the model should show the most important life criteria. Among these perhaps, the homeostasis is one of the most important ones. Homeostasis means that the system does not leave its stationary state, if the input parameters, i.e. the concentrations of incoming chemicals, the temperature of the environment, the osmotic pressure for transport of different chemical molecules through the membranes, etc., change even to a considerable extent. One should mention that the biological concept of homeostasis corresponds to the stability of the mathematical solutions of the system of equations. If the compartmentalized model cell does not show homeostasis, one has to assume (1) that the system of selected reactions is not large enough or does not contain, from the point of view of model-building, some of the most important reactions, (2) possibly the choice of compartments (spatial regions within the cell) was wrong and/or (3) the used reaction rate- and diffusion constants were unrealistic. In such a case, one has to improve the model by correcting one, two or all the three sources of errors, until the model cell shows homeostasis.

To illustrate the idea of compartmentalization of the reaction within the cell, let us choose first a very simple model system with 3 compartments and 10 different compounds. We assume that only compartment I interacts with the surrounding chemical bath, and three components (A, B and E) enter from it to this compartment. At the same time, two components (C and J) leave the system. The compartments are coupled by the common reactants D (in I and II) and F (compartments II and III) as well as through the dependence of the reaction rates and diffusion coefficients on the "nonactive" chemical components in each compartment.

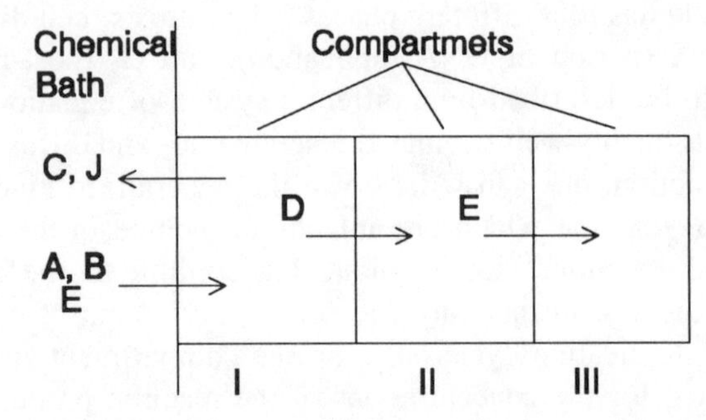

Fig. 6.1: A model system of three compartments and ten reactants. Components A, B and E enter the system from the surrounding chemical bath, and components C and J leave it (indicated by *arrows*). Component D takes part in the reactions in both compartments I and II, and component F takes part in the reactions in compartments II and III, respectively

Such a reaction scheme can be written down, for instance, as follows:

$$A + B \xrightarrow{k_1} 2C$$

$$2A + B \xrightarrow{k_2} D \qquad \text{Compartment I}$$

$$B \xrightarrow{k_3} C$$

in which the rate constants of the three reactions and all the diffusion coefficients in compartment I, between compartments I and II and between I and the bath, can be functions of components E, F, G, H, I, and J with the exception of the inward diffusion of E from the bath to I and the outward diffusion of J from I to the bath.

In a similar way one can write for compartment II

$$\begin{aligned} 2D + E &\xrightarrow{k_4} F \\ E + F &\xrightarrow{k_5} 2G \\ G + E &\xrightarrow{k_6} 2F \\ F + G &\xrightarrow{k_7} E \end{aligned} \qquad \text{Compartment II}$$

From the chemical reaction equations one can see that component D occurs here, as it did in I. The rate constants and diffusion coefficients can depend now on the concentrations of components A, B, C, H, I, and J.

Finally, one can have for compartment III, for instance, the reactions

$$\begin{aligned} 2F &\xrightarrow{k_8} H \\ 2H &\xrightarrow{k_9} I \\ A + B &\xrightarrow{k_{10}} 2J \end{aligned} \qquad \text{Compartment III}$$

which means that component F is active in both compartments II and III, and component J does not take part in any of the reactions. The same is true for component C produced in I. The rate and diffusion constants in the compartments can again be functions of the nonactive components A, B, C, D, E, and G.

To solve the kinetic equations for such a model system one can choose the compartments as cubes with edges of a length of 100 Å, i.e. their volumes are 10^6 $Å^3$. Since the concentrations of the chemical compounds should not be a strong function of space, very probably it will be enough to choose a rough grid with 10 Å intervals for the coordinates x, y and z. On the other hand, if we assume a 5 minute duration for one phase of a cell (20 minutes is a good average for the cycle time of a cell in a multicellular organism), one should take a time grid of at least 50 points, i.e. the time interval would be 6 s. This grid is probably good enough for slow reactions, but certainly for some reactions a much finer time grid would be necessary (at least for a specific part of the total time interval of 5 min).

For the rate constants typical values for enzymatic reactions between organic molecules were assigned and values were chosen for the diffusion constants which are correct within one order of magnitude.

The compartments were taken in a linear arrangement (Fig. 6.1) each one being a cube of (100 x 100 x 100) $Å^3$. The chemical components entering from the chemical bath to, or leaving the system of 3 compartments to the chemical bath are indicated by arrows in Fig. 6.1.

After writing down all the 10 kinetic equations with diffusion [12], they were solved numerically taking advantage of the subdivision of the 10 reactions into 3 compartments. One had to apply only 3 steps to obtain a consistent solution for the concentration of all the 10 chemical components.

This model study has shown that the idea of division of a cell into compartments to solve the coupled equations works well and therefore one can expect that the method can be extended without larger difficulties to several hundred biochemical reactions. The extension of the investigations for such a case is in progress in Erlangen.

6.3 Model for a Cell with a Changed Stationary State (Initiation of the Cancerous State)

After constructing a satisfactory cell model in the way described in the previous section, one can assume that due to the activation of oncogenes or inactivation of antioncogenes which influence the enzymatic (catalytic) activity of proteins, the rate constants of some chemical reactions, which are important for the self-regulation of the cell will considerably increase or decrease, because of overproduction of some regulatory proteins or inactivation and loss of them, respectively.

After a careful analysis of the results (concentrations as functions of space and time) of the stationary state solutions of a normal cell, one will be able to identify its most important regulation parameters as a function of the input concentrations, rate- and diffusion constants. One can then test the stability of the model against the change of rate constants and against the effects of external influences (toxic substances, drugs, etc.). In this way one can find out the critical values of the rate constant changes of the most important regulatory reactions at which the cell goes over to another stationary state. Furthermore, on the basis of the changed concentrations in this new state of the cell, one can characterize the new stationary state which is already the first step towards the malignant transformation. This procedure will provide a mathematical-physicochemical definition of a cancer initiated state of a cell.

Later the whole described procedure could be applied also to an ensemble of normal cells, including their interaction parameters in the mathematical procedure, and one can try to define through a systematic change of both the intracell- and intercell regulatory parameters the cancer-initiated stationary state of the cell ensemble under consideration.

Finally one should mention that Düchting [13] has developed a macroscopic electric regulation model-cycle for normal cell ensembles, for benevolent tumors and for malignant tumors. He can describe all the three states with the same regulatory cycle by assigning to the three different states different ranges of the electric parameters occurring in the cycle. He also gives criteria in which way the different electric elements in his regulatory cycle have to be changed to return from the, from a regulatory point of view, instable cancerous state to the stable normal state. One can postulate that with time his conclusions, based on a macroscopic electric regulatory cycle, can be compared and possibly justified on the basis of results obtained with our cell ensemble model, based on a physicochemical (chemical kinetics + diffusion) description of the normal and cancer-initiated stationary state. For the details of Düchting's model, since it contains a somewhat complicated mathematical formulation of his electric regulatory cycle, we refer to his original publication [13].

References

1. M.R. Pincus, J. van Ranousoude, J.B. Harfort, E.H. Chang, R.P. Carty and R.D. Clausner, Proc. Natl. Ac. Sci. USA **80**, 5253 (1983).
2. I.B. Weinstein (personal communication, 1992).
3. M. Eigen and P. Schuster, The Hypercycle, Springer, Berlin-Heidelberg-New York, 1979.
4. J. Ladik and M. Seel, Int. J. Quant. Chem. **QBS12**, 235 (1986).
5. P. Glansdorff and I. Prigogine, Thermodynamic Theory of Structure, Stability and Fluctuations, Wiley, New York-London, 1971.
6. G. Nicolis and I. Prigogine, Self-Organization in Nonequilibrium Systems, Wiley, New York-London, 1971.
7. J. Ladik and K. Laki, Int. J. Quant. Chem. **20**, 683 (1981); Y.-J. Ye and J. Ladik, Phys. Rev. **B48**, 5120 (1993); Y.-J. Ye and J. Ladik, Int. J. Quant. Chem., Proc. of the I. Congress of the International Society for Theoretical Chemical Physics (in press).
8. J. Ladik, Int. J. Quant. Chem. **QBS3**, 237 (1976).
9. See for instance: K. Haken, Synergetics, Springer, Berlin-Heidelberg-New York, 1977.
10. G.R. Welch, Progr. Biophys. Mol. Biol. **32**, 103 (1978).
11. See for instance: A.L. Lehninger, Biochemistry, Worth Publ. Co., New York, 1975.
12. U. Salzner, P. Otto and J. Ladik, J. Comp. Chem. **11**, 194 (1990).
13. W. Düchting, Kybernetik **5**, 70 (1968; in German with an English summary).

7 The Role of the Central Nervous System in the Malignant Transformation of Cells in Higher Organisms

It has been known for a long time that people who are frequently depressed, or who have suffered a great loss which affects them strongly emotionally (for instance the loss of a spouse) more frequently develop cancerous tumors than the average population [1]. This is usually explained by the fact that in the case of people with certain unhealthy states of mind, the material structure of the brain is also influenced (mind-brain parallelism). Since parts of the brain control the synthesis of the steroid type cancer promoters like phorbol esters [2] (see also Chapter 3), changes in the brain influence the synthesis and concentrations of these compounds.

It is well known, that in order to initiate cancer in an animal one has to expose it first to chemical carcinogens (in a low or medium concentration) and afterwards to a second type of compound, the so-called promoters. On one hand, the promoters, which never cause cancer by themselves, do not bind to the cell nucleus, but to the cell membrane and cause the segregation of a messenger molecule which affects DNA at the cell nucleus [3]. Note, that the action of promoters is usually reversible, unless the carcinoma has already developed (see Chapter 3, p.71). On the other hand, if one applies a large concentration of a strong chemical carcinogen the malignant transformation can take place without the promoters [3]. The work of v. Metzler may provide an explanation of these somewhat unusual facts. She has shown that if one gives a rat a large dose of 3-methylcholanthrene

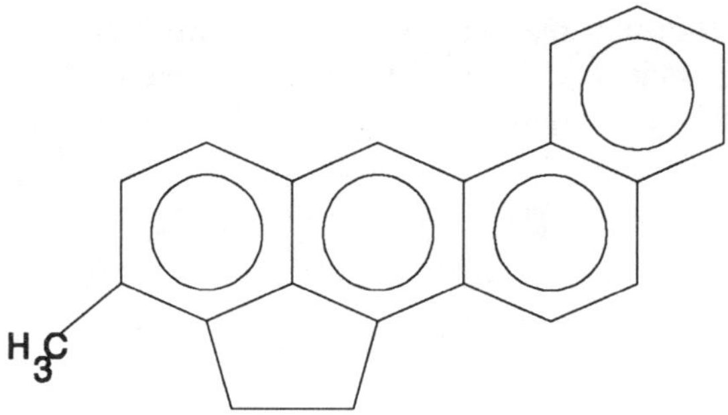

(which is one of the strongest carcinogens) or transplant a tumor to it, a significant change to the electroencephalogram (EEG) occurs. In the visual EEG spectrum a significant increase of the frequencies and of the amplitudes of the peaks between 3 and 18 Hz is observed. A mathematical analysis of the EEG curves has further proven this effect [4]. These changes occur within 24 hours before a chemical test at the site of the injection of the carcinogens shows any change. This change in the EEG spectrum persists during the latent period of a few months and also after the development of tumors until the death of the animal. One should mention in this connection that it was not checked whether the onset of other deseases also causes changes in the EEG spectrum and if so, whether these changes are desease specific (cancer specific, bacterial or viral infection specific, etc.). Despite the lack of these investigations, the observed facts reported by v. Metzler in the case of cancer seem to be interesting and important enough to the authors to report them at least in this very short chapter.

When the tumors have already developed, one can also find morphological changes in the brain (in the hypothalamus) of the animal [5]. It has also been found in a large number of experiments that if a drug acting on the central nervous system (CNS), like piracetam

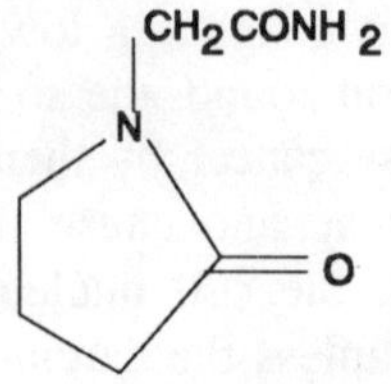

CH_2CONH_2

N

O

or pyrithioxine ($C_{16}H_{20}N_2O_4S_2$) etc. is administered simultaneously with the injection of the carcinogen, no changes in the EEG occur and later no tumor develops [6]. If piracetam is given to rats already injected with 3-methylcholanthrene but before the occurrence of tumors, it delays the appearance of the tumors [7]. It was further discovered that six days after the injection of carcinogens or tranplantation of tumors, changes are also observed in the concentrations of neurotransmitters like γ-aminobutyric acid (GABA)

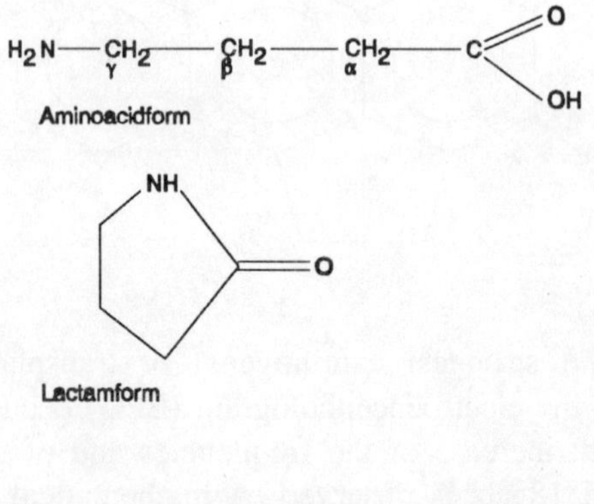

$H_2N—CH_2(\gamma)—CH_2(\beta)—CH_2(\alpha)—C(=O)OH$

Aminoacidform

NH

O

Lactamform

in different areas of the brain (in the hypothalamus and the hypocampus) and in the levels of different monoamines (like chatecholamine) which can be found both in the brain and in the periphery [8].

Finally it was detected that if one reinstates with different drugs acting on the central nervous system (CNS) the original concentrations of GABA or of other neurotransmitters, tumors can be cured [9]. This method of treatment for breast tumors has already been applied in a hospital in Frankfurt (M), FRG.

From all these facts, it is obvious that there is a strong relation between local tumor development and the CNS in higher organisms [10]. One can make the new assumption that if a larger dose of a potent carcinogen is given to a test animal and the carcinogens influence the brain as described above, these changes in the CNS influence the synthesis of different steroid type compounds which are not snthesized in the brain, but their synthesis is controlled by the brain. On the other hand, it is known that certain steroid type compounds act as promoters in carcinogenesis (see above). This means that strong carcinogens in larger doses are able to increase the synthesis of promoters and therefore the promoters would not have to be added from outside to initiate a malignant transformation of the cell. This strong interaction between the brain and the peripheral regions where the original attack of the carcinogen occured seems to be unusual at first sight, but if one takes into account that in higher organisms all functions, including sensory perceptions, motions, etc., of the whole body are completely controlled by the CNS, the experimental findings described here are not surprising at all.

One arrives in this way at a unified hypothesis for the role of the CNS in carcinogenesis. If strong carcinogens in large concentrations are injected into a test animal, enough carcinogens reach the brain to depolarize the membranes of enough brain cells to manifest themselves in significant changes in the EEG spectrum. This then causes the increase of the GABA content in the hypothalamus and the hypocampus and a reduction of the concentrations of monoamines or their metabolites in the hypothalamus and the caudate nucleus. All these changes subsequently cause morphological changes in the described parts of the brain. The latter can be observed experimentally only after the tumor has developed at the primary site of carcinogen injection, but most probably occurs much earlier to a smaller extent (which cannot be experimentally detected yet) even before the occurrence of the tumor. These unobservable, but in all probability already existent electric, chemical and morphological changes in the brain at the early stage of carcinogenesis can then increase the synthesis of different steroids which act as tumor promoters. In other words, a strong carcinogen in a larger concentration facilitates via the brain the synthesis of tumor promoters. This is most probably the reason why in this case no extra administration of promoters is needed for tumor development.

On the other hand, if drugs acting on the CNS are given to a test animal simultaneously with the carcinogen injection, they most probably hinder the depolarization of the membranes of brain cells by the carcinogen, and thus the described chain of events does not take place.

Finally in the case of an increase of the concentration of GABA due to carcinogenesis, its reduction by pyrithioxine or piracetam to its normal level in the brain (which decreases the rate of tumor growth) most probably counteracts the increase of cancer promoter synthesis. The same can be said about the increase of the concentration of catecholamine to its normal level (it is reduced due to the action of carcinogens in the brain) through adrenaline:

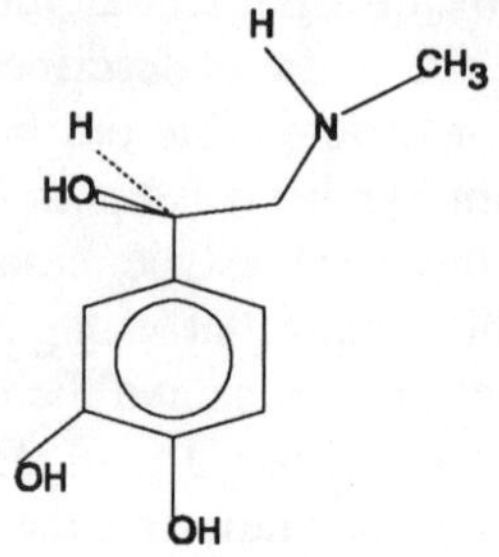

dopamine:

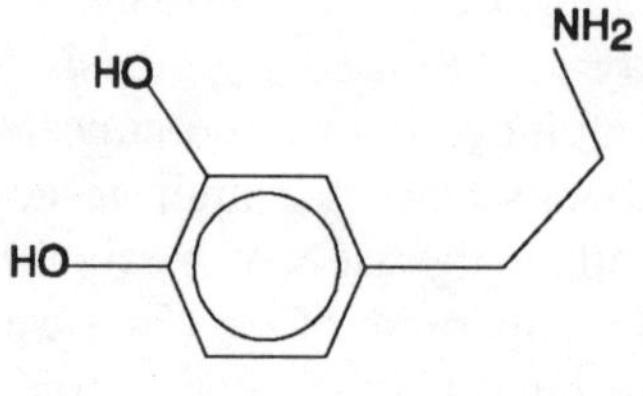

levodopa:

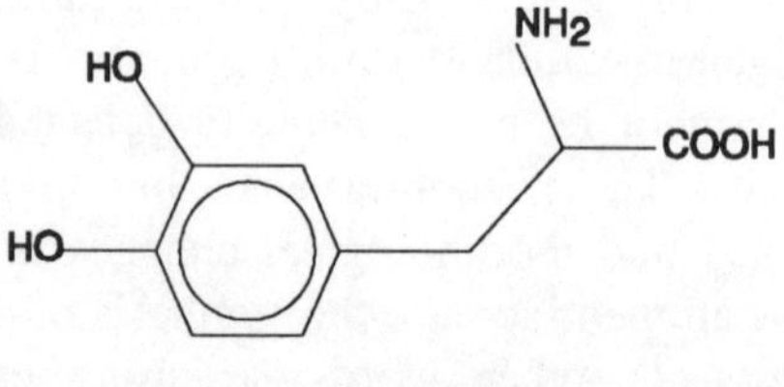

and imipramine (10,11-dihydro-*N*,*N*-dimethyl-5*H*-dibenz[b,f]azepine-5-propanamine) [11].

Of course, we can be certain only about the experimentally established facts. To prove their interpretation and the relations between these events, assumed above, a great deal of experimental work is still needed both on brain research and on cancer research. In the latter case research is especially needed about the role of tumor promoters during tumor growth.

References

1. M. Blohmke, Institute of Medicine of Labor, University of Heidelberg, personal communication, 1977.
2. See for instance: I. B. Weinstein, H. Yamasaki, M. Wigler, L.-S. Lee, P. B. Fisher, A. Jeffrey, and D. Grünberger, in "Carcinogens: Identification and Mechanisms of Action", (eds. C. Griffin and C. P. Show) Raven Press, New York, 1979, p. 399; I. B. Weinstein, J. Arcolo, M. Lambert, W. Hsiao, S. Gattoni-Celli, A. M. Jeffrey, and P. Kirschmeier, in "Molecular Biology of Tumor Cells", (eds. B. Wahren et al.) Raven Press, New York, 1985, p. 55.
3. I. B. Weinstein, personal communication, 1986.
4. A. v. Metzler, Z. Krebsforschung **83**, 195 (1975) (in German).
5. W. R. Brain, P. M. Daniel and J. G. Greenfield, J. Neurol. Neurosurg. Psychiatry **14**, 59 (1951); A. V. Metzler and C. Nitsch, Cancer Detection and Prevention **9**, 259 (1986).
6. A. v. Metzler, J. Cancer Res. Clin. Oncol. **94**, 630 (1979); ibid. **95**, 11 (1979); A. v. Metzler, Z. Krebsforschung **77**, 300 (1972) (in German); A. v. Metzler and C. Nitsch, Naturwissenschaften **69**, 48 (1982).
7. A. v. Metzler and C. Nitsch, J. Cancer Res. Clin. Oncol. **101**, 339 (1981).
8. See [7] and A. v. Metzler and C. Nitsch, Naturwissenschaften **69**, 48 (1982).
9. A. v. Metzler, personal communication.
10. A. v. Metzler and C. Nitsch, Cancer Detection and Prevention **9**, 259 (1986); J. Ladik, Physiol. Chem. and Phys. and Med. NMR **20**, 87 (1988).
11. A. v. Metzler and C. Nitsch, Cancer Detection and Prevention **9**, 259 (1986).

8 What Should Be Done

8.1 Introduction

Apart from the standard methods in biomedical cancer therapy, such as surgery if possible, followed by X-ray radiation treatment and chemotherapy repeated at intervals, the question arises what else could be done. This question is especially topical in face of the facts, that although there have been improvements in early detection and developments in the three classical methods of tumor treatment and the probability of cure has been significantly increased, as in case of breast cancer, cancer of the intestines, leukemia, certain types of skin cancer, etc., the overall five-year survival rate of all cancer patients has improved by less than ten percent in the last thirty years. The main reason for this is, that the most frequent and dangereous tumors develop mostly inside the body at places where they cannot be easily operated, as it is the case for e.g. brain-, lung-, liver-, stomach-, lymphoma-, pancreas tumors, etc..

In the case of an imperfect operation, the radiation and subsequent chemical treatments do not help very much, especially since both of them are not strongly tumor cell specific. To be objective, for an explanation of the slow increase of the five years survival rate, which is the probability that the patient is alive five years after the detection of the primary tumor, one has to take into account two factors which make the statistics worse: (1) the increase in the average life expectation, since the probability of tumor development increases in humans with the fourth power of the age, and (2) the increase of chemical and other possible pollution (electrosmog?). In the case of the non or not easily operable tumors a further aggravating factor is that, since they are mostly situated inside the body, their early detection is also more difficult.

In which direction should research go in this situation to supplement and improve the unquestionably very useful and necessary methods for the treatment of tumors generally used in medicine? In the opinion of the authors, besides further improving the methods which are available, the development should be threefold: (1) make the curative methods more cancer cell specific, (2) improve further and apply more widely the methods for early detection and last but not least (3) work out new and better methods for the prevention of tumor development. The first two points will be discussed here only shortly, because they do not belong to the scope of this book, but the question of new methods will be discussed in more detail.

8.2 Some Cancer-Cell-Specific Curative Methods

It is well known that cancer cells have different cancer antigens on their surface than the corresponding normal ones, though in some cases they are masked by a specific protein [1]. It was possible to clone the antibodies of certain antigens (monoclonal antibodies) and to bind cytostatica to them. It was hoped that in this way the drugs acting against certain types of cancer would reach specifically only the tumor cells. There are, however, two difficulties with this method. First of all it was not always possible to find the cancer cell specific antigens, especially when they were masked, and therefore the corresponding antibodies could not be cloned. A second quite general difficulty is that even if the right cytostatica are bound to the right monoclonal antibody, the cytostatica, in this way, mostly only reach the cells in the outer region of the tumor and cannot easily penetrate inside it. This is of course very much dependent on the consistency of the tumor, which, in turn, depends strongly on the kind of tumor and on the type of the normal organ in which the tumor has developed. For instance in the case of non-Hodgkin lymphoma sarcoma, of which there are about 50 different types, the treatment using cytostatica bound to these cancer-specific monoclonal antibodies was successful only in 4-5 cases [2]. It is obvious that substantially more biochemical and immunological research is needed to use the full potential of cancer cell specific monoclonal antibodies.

The other potentially cancer-cell-specific treatment is their irradiation with π mesons. π mesons have a rather short lifetime after which they decay into μ mesons. Their lifetime, however, is dependent on their velocity with respect to the object which is bombarded by them. According to the theory of relativity their lifetime increases if their velocity increases. π meson beams can be generated with different energies (velocities) and thus their lifetime can be tuned. If one knows the location of a tumor in the body, one can choose the distance of the π meson source and the velocity of them in such a way, that the π mesons should decay exactly inside the tumor. Namely, if a π meson decays exactly at the site of the tumor according to the laws of physics it should have a much larger effect there, than in the tissues which it passes on its way to the tumor. According to the information available to the authors, attempts to apply this method have not worked out well up to now [3]. Obviously further strongly interdisciplinary cooperation between physicists and scientists in biomedical research is necessary to develop the idea sketched above to a practical, easily applicable tumor-cell-specific method.

8.3 Developments in Early Tumor Detection

In Chapter 7, the connection between tumor induction and development, respectively, on the one hand and changes in the EEG spectra of the test animals on the other hand were described. These changes in the electroencephalograms could also be used for the early detection of tumors in humans. For this, it should be better established, how tumor specific these changes are. Further one should find out whether these changes are specific for the type of tumor and for the organ in which it develops. All this, again requires further interdisciplinary research.

It is known that nuclear magnetic resonance (NMR) computer tomography is also widely used to locate the early development of a tumor in the body. It is also known that there is world wide haematological-biochemical research for a blood test which should indicate the development of a malignant tumor in its early stages. Certainly these techniques will be developed further to a large extent.

Finally we should like to mention the method of thermoregulation which is rather well established but not widely used. Namely the body emits infrared (IR) radiation with a spectrum (a distribution of emission peaks with different frequencies and intensities), which is characteristic for certain parts (organs). If in a certain organ a tumor starts to develop, the spectrum of the characteristic infrared radiation changes.

This static method has been generalized to a dynamic, i.e. time dependent form [5]. Namely one can put one of the patient's hands into ice-cold water for a few minutes. In this case all the spectra of the organ specific infrared radiations change. After taking out the hand of the patient of the cold water, the IR spectra emitted by the different organs return to their original forms with different time delays, the so-called relaxation times. On the other hand if in one organ a tumor is developing, its spectrum will return to its original form with a longer relaxation time. This dynamic form of the method, from which the name thermoregulation comes, is more sensitive than the static one [5]. For this reason the development of a tumor in an organ can be detected with the help of this method earlier than with its static version.

8.4 Possibilities of Prevention of Cancer

There are trivial and well known methods for the prevention of cancer, e.g. refraining from smoking, avoiding passive smoking, and restricted use of alcohol which weakens the immune system. Further one should follow a sensible diet, e.g. avoiding smoked meat or fish or plant products, which are known to contain chemical carcinogens like coconut or plants which were grown by using carcinogen containing herbicides. One should also avoid areas with strongly polluted air, strengthen the immune system with certain proteins like interferon, etc..

At this point we will discuss in more detail further methods for cancer prevention based on the theoretical and experimental results described in Chapters 3 and 5-7. In Chapter 3, a concise review was given about the most important chemical carcinogens (initiators) and their reactions in the cell which lead to their ultimate oncogene-activating forms. If the chain of reactions of a given carcinogen in the body is known, then with the help of appropriate biochemical research it could be found out how some of the crucial reaction steps on the way to its ultimate form could be hindered by reacting it with appropriate chemicals. This kind of approach has been pursued [6], but there are two major difficulties. First of all one has to know from epidemiological studies to which carcinogens a certain segment of the population of a given region is primarily exposed, since one cannot hinder the reactions of all possible chemical carcinogens simultaneously. A still larger difficulty arises from the fact, that usually if a given drug hinders certain chemical reactions of a primary carcinogen on its way to its ultimate form, the very same drug may also hinder with a high probability other biochemical reactions vital for the normal functioning of a number of types of cells in the organism. One can hope, however, that with the help of well focused biochemical research one can find certain molecules which hinder a crucial step in the formation of the ultimates of certain carcinogens without much interference with normal cell functioning.

The second line of defense (prevention) is to find out, which are, among the theoretically possible ones, the most important long-range mechanisms of oncogene activation by chemical carcinogens and the mechanisms by which radiation causes double strand breaks in DNA, connected with a possible loss of antioncogenes (see Chapter 5). This would require a rather intensive and extensive experimental research on the solid state physical properties of the biopolymers, first of all of DNA and proteins and on their interactions.

The main difficulty of this kind of solid state physical measurement on biopolymers is the absence of samples which are adequate for such experiments. The DNA or protein samples, produced either by biochemical isolation, genetic engineering methods or in vitro synthesis, usually contain 1-5 per cent unknown impurities. This makes it impossible to perform reliable measurements of their transport properties, e.g. metallic or hopping electric conductivity (see Appendix) and mobility measurements, Hall effect and heat conductivity measurements, etc.. Further their electric and magnetic properties or even of their phonon spectra among others cannot be measured with available samples.

To overcome this formidable difficulty, which is partially an objective one and partially also a psychological one on the part of the community of scientists in biomedical research, which is first of all due to their not very strong background in physical chemistry and physics, one needs first of all a serious so-called biomaterial science development. This means that with the combination of different modern purification techniques, like different forms of high performance liquid chromatography (HPLC) [7], paper electrophoresis, ultracentrifugation, etc., most of the occasional impurities should be removed from the samples so that the amount of non-structural impurities should be less than 0.1 per cent. After purification, the samples have to be characterized. That means that the amounts and types of the remaining inorganic and organic structural and non-structural impurities have to be determined as well as possible. This could be done relatively easily for the remaining inorganic impurities with the help of laser emission spectroscopy. In this method, the sample is burned by laser light in which case the emission spectra (intensities and frequencies) of all inorganic ions in the sample can be measured.

The organic impurities and their concentrations, if they are fluorescent, can be identified with the help of the fluorescent line narrowing technique, induced by a monochromatic laser [8]. If the organic impurities are not fluorescent, their determination becomes more difficult, but with the help of the combination of different spectroscopic methods, like electronic, vibrational and resonance spectroscopy, certainly successful new methods could be worked out to find out their identities and amounts.

To characterize the biopolymer samples adequately, one has to know their sequences and conformations, and further the lengths of the different chains in a sample should be approximately the same. To determine, whether one has really obtained a more or less homogeneous sample with respect to chain length, one can use, among other methods, the small angle neutron scattering techniques [9].

We are perfectly aware of the fact, that the development of biomaterial science described here requires a major research effort. On the other hand, if it were successful it would mean a breakthrough in biological sciences generally and in experimental biophysics in particular. Such samples applicable for solid state physical measurements on biopolymers would not only allow the determination of the most important long-range mechanisms of oncogene activation by chemical carcinogens and of antioncogene inactivation via double strand breaks by radiation, but also facilitate a completely new point of view of biopolymers as complicated solids and not as an ensemble of interacting subunits (molecules) as most biologists, chemists and even many quantum chemists nowadays look at them. This new view will help us a great deal to generally understand radiation damage, mechanisms of viral infection, enzyme activity, transfer of genetic information, charge, energy and signal transport in the cell, to mention only a few of the basic problems.

Returning to our problem - the initiation of cancer - after the proposed biomaterial science development, it could be decided which are really the most important long-range mechanisms in DNA which are involved in cancer initiation.

Besides the mechanisms mentioned in Chapter 5, one can also think of vibrational energy transfer along a nucleotide base pair stack, on propagation of a concentration change of K^+ ions caused by the binding of a carcinogen to DNA and on many other still more complicated collective states. The occurrence of such states could be always triggered by attacks of chemical carcinogens or by direct hits of electromagnetic or particle radiation.

After understanding the basic solid state physical mechanisms of oncogene activations or antioncogene inactivations caused by external agents, there exists a fair possibility that one might be able to find out combined chemical and physical methods which can counteract these events. Such new methods which can be worked out only with the close cooperation of medical doctors, biochemists, chemists, and experimental and theoretical physicists would have a rather good chance of preventing the initiation of cancer caused by chemical carcinogens or radiation.

If the initiation of cancer cannot be stopped at the DNA plus carcinogen level, and therefore the self-regulation of the cell becomes disturbed there is still a third chance of interrupting the chain of events which leads to the malignant transformation of a cell. First of all as we have seen in Chapter 7 the administration of drugs acting on the central nervous system, such as piracetam or pyrithioxin can reduce the concentration of the neurotransmitter GABA, which increases after cancer initiation. Likewise, other drugs can increase the carcinogen-caused excessively low concentration of the monoamine-type neurotransmitters. Of course such a treatment should be applied only if there is a suspicion of the occurrence of a precancerous state for an ensemble of cells at a certain point in the body.

The same can be said about a chemical inactivation of the oncoproteins coded by a given oncogene in a cell ensemble. To establish the existence and location of a precancerous state somewhere in the body, very probably the above described thermoregulation technique could be used. It is logical to assume that there is a continous change in the IR spectrum of cells of certain types when they are transformed from their normal state through different stages of precancerous states, to their final malignant state. To be able to find out whether this is really the case and to characterize these different IR spectra, a large number of further experiments is still needed.

In the case where there is a suspicion that an individual has suffered a high dose of radiation at a certain part of the body and that this could have lead to the loss of a part of the antioncogenes, one could again use the thermoregulation technique to discover whether the self-regulation of a certain cell ensemble has been disturbed or not. If the latter is the case, and more would be known about the most important antioncogenes and antioncoproteins in certain types of cells, one could supply the missing antioncoproteins from outside and in this way re-establish the normal stationary state of the cells under consideration.

One should mention that the maintenance and disturbance of the self-regulation of a cell certainly involves a large number of regulatory proteins. Therefore both experimental and theoretical (see Chapter 6) studies are needed to establish which enzymes are involved in a certain perturbation of the self-regulation of a certain type of cells. The theoretical investigations should be extended, starting from the cell model described in Chapter 6, to include interactions of model cells, which in a malignant state should also show the lack of contact inhibition, an experimentally well-known fact. Finally this should lead to a general regulation theoretical investigation of a cell and of an ensemble of cells to find out in detail, how oncoproteins or the lack of antioncoproteins disturb the self-regulation of a single cell or of an ensemble of cells. The "inactivation of oncoproteins" or "supply of antioncoproteins from outside", as described in the previous paragraphs should be understood in this more general regulation theoretical sense.

To conclude, one can say that the most important initial steps of carcinogenesis and their possible prevention could be understood on the basis of the scheme shown in Fig. 8.1.

The abbreviations in Fig. 8.1 at the different boxes are:

1. Chemical Carcinogens,
2. Ultimate Carcinogens,
3. Binding to DNA,
4. Oncogene activation, Oncoprotein Overproduction,
5. Disturbance of Cell Self-Regulation,
6. External Promoters,
7. Brain,
8. Promoter concentration changes caused by changes in the brain,
9. Promoter Binding to Cell Membrane,
10. Production of Messenger Substances produced because of promoter binding to the cell membrane,
11. Direct Hit of Radiation on DNA,
12. Double Strand Breakings of the DNA double helix caused by direct hits of radiation,
13. Inactivation of Antioncogenes caused by the loss of genetic information due to DSB's in DNA, Lack of Production of Antioncoproteins (regulatory proteins which suppress DNA duplication).

This figure of course cannot indicate the detailed, at present only partially known mechanisms of each step, indicated by arrows, but tries to summarize the network of subsequent events as described in Chapters 3 and 5-7 of this book. The double arrows show the possible points of intervention, as described in this Chapter, the combination of them would most probably prevent the start of the malignant change of a cell.

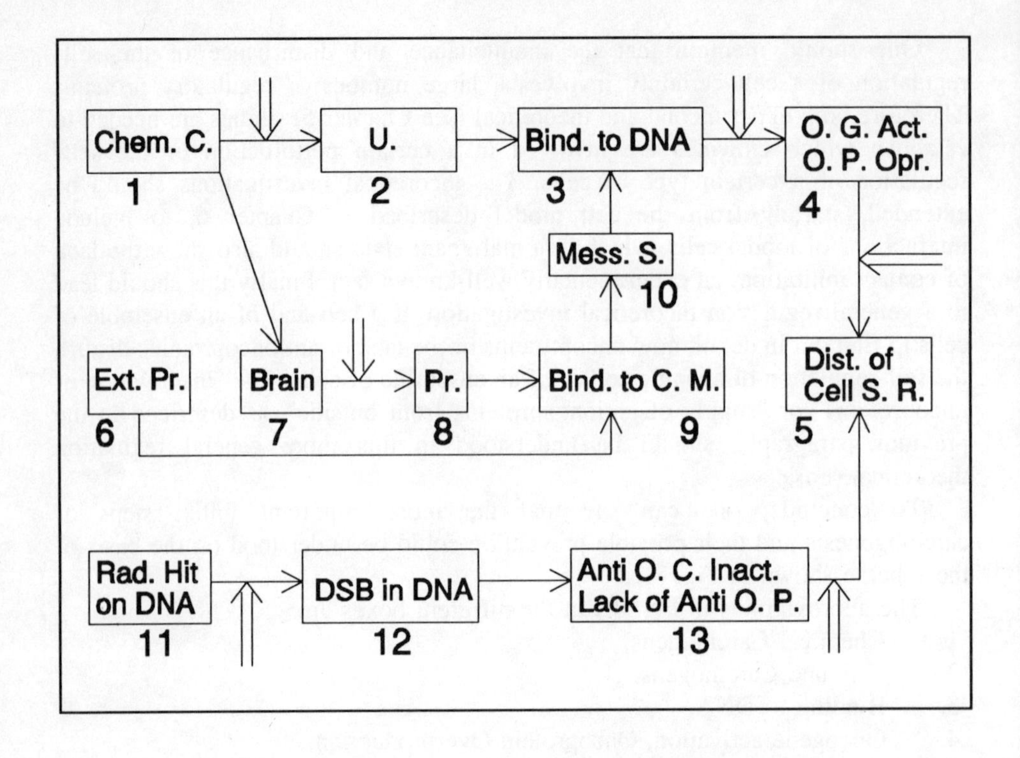

Fig. 8.1: The network of events which most probably lead to the initiation of cancer in a cell together with the steps where with appropriate intervention the process could be hindered (prevention of cancer). The *arrows* indicate the relations of subsequent events, the *double arrows* possible points of interventions as described in the text above to prevent the initiation of the malignant transformation of the cell

References

1. K. Laki, personal communication (1976).
2. V. Devita, personal communication (1991).
3. Z. Fuks, personal communication (1991).
4. M. Blohmke and G. Heim, Erfahrungsheilkunde **29**, 553 (1980), in German.
5. M. Blohmke, G. Heim, H.-P. Stof, and G. A. Bothmann, Phys. Med. u. Reh. **21**, 548 (1980); G. Heim, M. Blohmke, G. A. Bothmann, and H.-P.

Stof, Med. Klin. **76**, 108 (1981); M. Blohmke, G. Heim and H.-P. Stof, Erfahrungsheilkunde **31**, 573 (1982); M. Blohmke, G. Heim, W. Koenig, and H.-P. Stof, Krebsgeschehen **5**, 123 (1982), all in German.
6. I. B. Weinstein, personal communication (1990).
7. See for instance: A. M. Guichon and M. Martin, J. Chromatography **3**, 326 (1985).
8. See for instance: E. L. Wehry and G. Mamantov, in "Modern Fluorescence Spectroscopy", Vol. 4, E. L. Wehry ed., Chapter 6 (1981).
9. See for instance: R. M. Moon, Science **230**, 274 (1985).

Appendix: Outlines of a More General Theory of Cancer Initiation in the Cell

A.1 Introductory Remarks

In the main part of this book we basically followed the generally held opinion about cancer initiation: external factors (chemical carcinogens plus promoters as well as radiation) activate oncogenes or inactivate antioncogenes in DNA. According to the central dogma of molecular biology these changes in the regulation of (anti)oncogenes influence, through the corresponding messsenger RNA molecules, the amounts of the proteins which are coded by these genes. Finally the overproduction of these so-called oncoproteins and the lack or the decrease of the production of some cell duplication suppressor proteins, the so-called antioncoproteins, leads finally to the perturbation of the self-regulation of the cell. In this way, the cell, which is a thermodynamically open system, switches over to another stationary state with a different self-regulation. This can be interpreted as the initiation of a malignant transformation of the cell.

In the first chapters of this book we described briefly the biochemical knowledge about oncogenes, antioncogenes and their activation or inactivation, respectively. Furthermore, we summarized the known facts about the most important chemical carcinogens, including, in some cases, the description of the reactions inside the cell which lead to their biologically active forms, the so-called ultimates. This was followed by a brief description of different effects of electromagnetic and particle radiations if they directly hit DNA (Chapters 1-3). After that a short summary of the structure and biological functions of DNA and different proteins was given (Chapter 4).

After this preparation we went to the molecular and solid state physical treatment of the electronic structure of DNA and proteins in a qualitative form. On the basis of such very large scale calculations we proposed different long-range solid state physical mechanisms to understand in more detail, that is also in a physical sense, how binding of chemicals to DNA or direct hits of radiation on DNA can cause oncogene activation or antioncogene inactivation (Chapter 5). In Sect. 8.4, we also gave some suggestions how the proposed physical mechanisms could be experimentally verified or falsified. All these discussions strictly followed the central dogma: everything has to start with changes in DNA.

Now we can try to turn the tables and pose the question whether it is not possible to disturb the self-regulation of the cell and in this way initiate cancer if we change the chemical structure, including conformation, or physical state of certain proteins. For this purpose we shall describe qualitatively in Sect. A.2 some recent results which show that proteins, if they possess free charge carriers -

which in vivo can very easily occur with the help of charge transfer - are good disordered (amorphous) semiconductors with rather large hopping conductivities.

A.2 Proteins as Good Disordered Amorphous Semiconductors

Szent-Györgyi published two papers [1,2] in 1941 in which he postulated electronic conduction along the main polypeptide chain of a protein. He came to this idea first of all on the basis of the oxidation-reduction reaction cycle of oxygen metabolism, the so-called Szent-Györgyi-Krebs cycle [3], in which electrons are transferred through several different proteins. There is a similar situation in photosynthesis or generally in the case of oxidation-reduction enzymes.

There were both experimental [4] and theoretical [5] attempts to prove this hypothesis. The experiments in which a weak semiconduction was found in different proteins [4] didn't seem to be very reliable, because of the difficulties with the preparation of appropriate samples (see the discussion in Sect. 8.4). With these measurements it was even impossible to decide whether the conduction comes from the proteins or from impurities in them.

The theoretical calculations were mostly based on a periodic protein model with repeat units of the form

$$-\overset{\overset{\vdots}{O}}{\overset{\|}{C}}-\underset{\underset{\vdots}{H}}{\underset{|}{N}}-$$

Thus there are crystal orbitals through the hydrogen bonds perpendicular to the main polypeptide chain (see Fig. 5.14 in Chapter 5). These semiempirical band structure calculations have provided a too small gap around 3 eV, while the correct value is about 7 eV, and the band widths were too small due to the weak coupling of the units through the hydrogen bonds. Brillouin [6] assumed in 1962 that the conduction occurs along the main polypeptide chains and that the different side chains act as "impurities". It was, however, still impossible to prove the correctness even of this model because the theory of non-periodic (disordered) chains was not developed enough at that time and the available computational facilities were not sufficient to perform the necessary complicated calculations. For all these reasons, the scientific community did not accept Szent-Györgyi's hypothesis, though he kept on emphasizing it in his books [7].

There were two main arguments against Szent-Györgyi's assumption: 1) the gap is too large to have free charge carriers and therefore no intrinsic conduction along the main chains of proteins is possible, and 2) a polypeptide chain built up of 20 different amino acid residues is highly disordered and therefore no conduction can occur along it.

To explain electron transport through proteins the following mechanisms were considered as important: 1) electron transport coupled with proton or ion transport within a protein and between different proteins, 2) electron hopping through different parts of a protein, which are close together in space due to the complicated folding of the protein, and 3) electron transport through tunneling among different parts of a protein and between different proteins.

Without denying the existence of the three mechanisms for electron transport through proteins mentioned, we were able to prove with the help of the modern theory of disordered chains, appropriately generalized in the case of polypeptides, that rather good semiconductivity occurs in pig insulin [8] and in hen egg white lysozyme [9] along their main chains if there are free charge carriers in these proteins. The problem of free charge carriers does not seem to be a serious one, because in vivo, a larger number of charge transfers always occurs. This is due to the interactions of the protein chains with other molecules or chains which give over or take away electrons. The proof of good semiconductivity through the hoppings of the electrons between different units of these two aperiodic proteins required a rather complicated mathematical formalism which we cannot describe here, but give for the interested reader the proper references [10-15]. One should also mention that these calculations which were performed by taking into account the experimental sequences and conformations in both cases required several hundred hours on a CRAY YMP supercomputer.

In Fig. A1 the calculated hopping conductivities $\sigma(\omega)$ in $\Omega^{-1}cm^{-1}$, more precisely the logarithm of their absolute values, as a function of the logarithm of the frequency ω (in s^{-1}) are shown. One can see that the insulin curve falls between those of good semiconducting amorphous (disordered) glasses, which show also a hopping mechanism. One should mention also that if one would use more accurate approximations in the calculations, according to our estimation the $|\sigma(\omega)|$ values would further increase by about half an order of magnitude at all ω-values, which is indicated by arrows on the insulin curve in the figure.

We have extended the calculations up to $\omega=10^{20}$ s^{-1} for insulin and also performed them for lysozyme. As one can see from Fig. A2 and Fig. A3, respectively, the $\log_{10}|\sigma(\omega)|$ curves become in both cases constant, as one would expect on the basis of theoretical considerations [9], at $\omega=10^{11}$ s^{-1} with a $|\sigma(\omega)|$ value of 10^{-2} $\Omega^{-1}cm^{-1}$ in the case of insulin and of 10^{-4} $\Omega^{-1}cm^{-1}$ for lysozyme, respectively. We intend to extend these calculations also to other polypeptide chains improving at several points also the approximations involved [9].

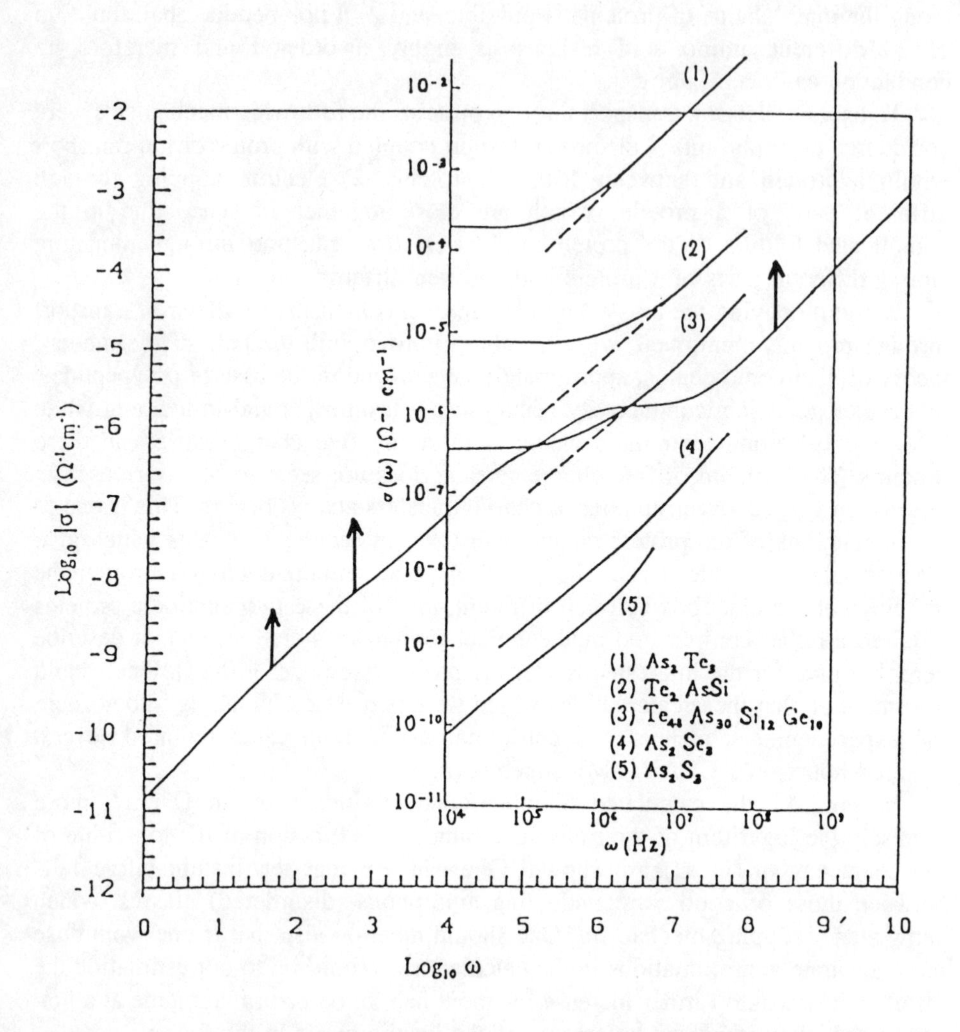

Fig. A1: Calculated logarithms of the absolute values of the conductivities ($\log_{10}|\sigma(\omega)|$) in units of $\Omega^{-1}cm^{-1}$ as a function of the logarithm of the frequency ω (in s^{-1}) for pig insulin (the curve goes from $\log_{10}\omega=0$ up to $\log_{10}\omega=10$). The other curves contain in the 10^4-10^8 s^{-1} frequency range the measured values of different chalcogenides, which are good semiconducting amorphous glasses [10]

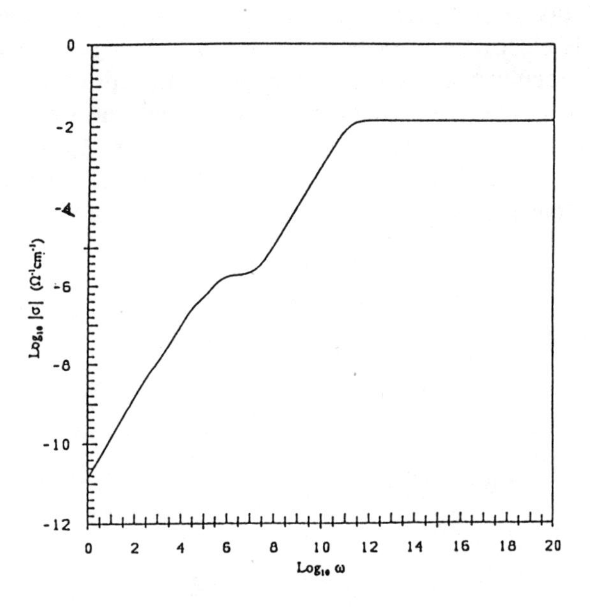

Fig. A2: The $\log_{10}|\sigma(\omega)|$ versus $\log_{10}\omega$ curve for insulin up to $\omega=10^{20}$ s^{-1}

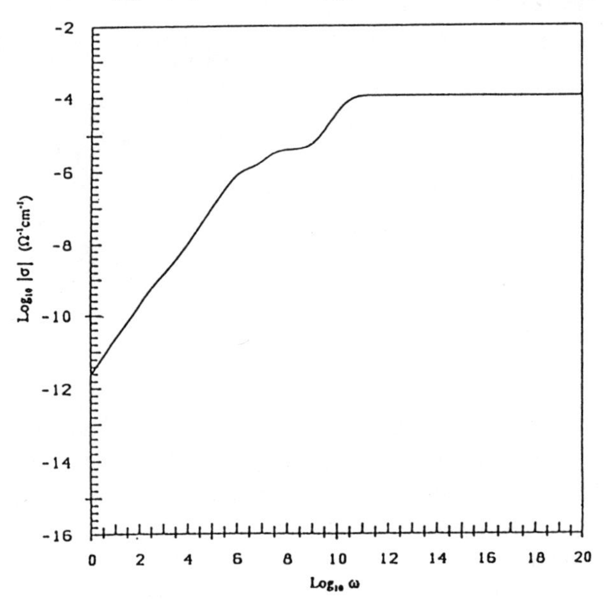

Fig. A3: The $\log_{10}|\sigma(\omega)|$ versus $\log_{10}\omega$ curve for lysozyme

Independent of the future work, one can safely state on the basis of the results already obtained, that Szent-Györgyi's conjecture of 1941 was correct: there can be an unnegligible electronic conductivity along the main polypeptide chains of proteins if certain segments of the main chain point in the direction of the electric field strength which causes the conduction, or are only at a small angle to it. Therefore the possibility of hopping electronic conduction along certain parts of a folded protein molecule has to be taken into account together with the other mechanisms mentioned above (see Fig. A4)

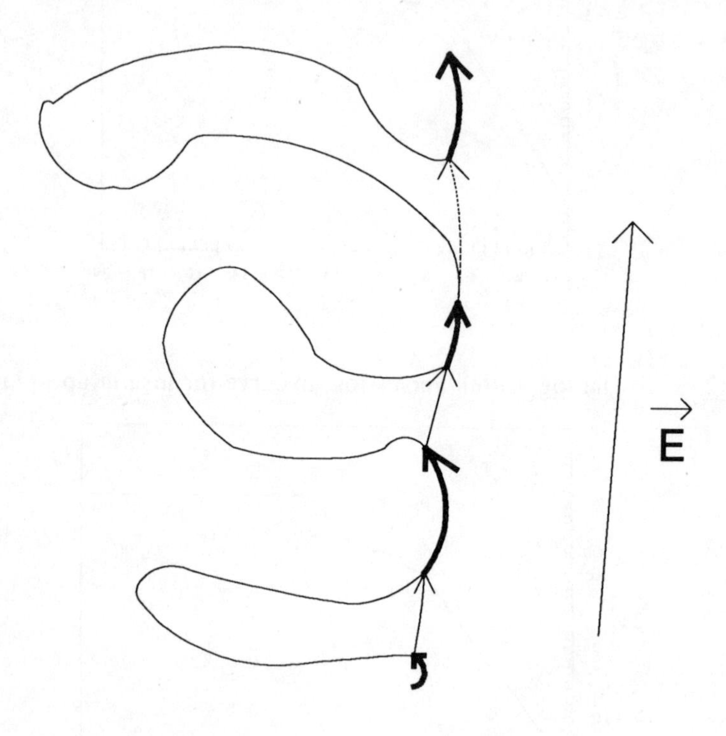

Fig. A4: Schematic representation of a folded protein chain. The *arrow* with the vector of the electric field of strength E denotes the direction of the field. The *thick segments with arrows* at their ends denote hopping conduction along segments of the main chain, the arrows between near lying turns of the chain indicate intersegment hopping and the *broken line with arrow* intersegment tunneling (schematic)

A.3 The Normal and Cancerous State of Living Matter

On the basis of his strong belief in electronic conduction in proteins Szent-Györgyi assumed that in eukaryotic (differentiated) cells more, or less unhindered electronic flow through proteins is necessary for their normal functioning [7]. Later he also postulated that if this electron flow in proteins is hindered, no oxygen metabolism but only fermentation is possible in the cell which becomes in this way cancerous see especially the third book of ref. [7].

This latter assumption agrees well with the experimental findings of Warburg [17] that if one suppresses the oxygen metabolism of a tissue culture consisting of eukaryotic cells, the cell culture changes its metabolism to fermentation (like cells of unicellular organisms) and becomes cancerous.

It is easy to imagine that the free electron flow (and with it energy and signal flow) in and between proteins facilitates very much the normal self-regulation of a cell. Therefore it is not difficult to visualize that if chemicals binding to proteins, or radiation damage suppresses this electron transport this may change the self-regulation of a cell into the direction of a cancerous state. The same can be said if binding of chemicals or radiation damage inactivates regulatory enzymes by changing their conformation or breaking bonds in them, especially at their active site.

For all these reasons, the proof of which would require further thorough biochemical and biophysical work on well defined samples (see Chapter 8), it seems to be plausible to assume that the initiation of a malignant transformation of the cell does not necessarily start via activation of oncogenes or inactivation of antioncogenes in DNA caused by chemical carcinogens and radiations, though this is probably the most effective way of cancer initiation. The very same external agents most probably can also attack regulatory enzymes directly with the same end result: perturbation of the self-regulation of the cell which brings it over into another (precancerous) state (cancer initiation). Of course DNA and regulatory enzymes can be attacked simultaneously by external carcinogens which is most probably the case in practice.

In the unprobable case that the malignant transformation starts alone by carcinogen attack on regulatory enzymes, the question arises, why after cell duplication are the daughter cells also cancerous. If we think very generally about the self-regulation of the cell, which also includes the regulation of the transcription process from DNA to RNA and the translation of the genetic information from messenger RNA molecules to proteins, one can see many ways how these processes are also deregulated. This can again cause the activation of oncogenes or inactivation of antioncogenes with the known consequences.

Although the biological information usually follows the route DNA → RNA → protein, changes in regulatory enzymes can easily cause a reversed process:

changes in certain proteins can cause changes in the regulation of transcription and translation processes originating at the DNA molecules.

If we look at a cell as a general self-regulatory system, one can easily assume that any change which disturbs this self-regulation, independently of whether the primary change happens at DNA, at enzymes or at other components of the regulatory system of a cell, can push it over to another stationary state. This state may be precancerous, lethal or another still normal state. Such a change of the state of the cell can act back on the regulation of its DNA molecules which makes the change hereditary. Of course to prove this perhaps somewhat far-fetched hypothesis a large amount of theoretical work for a better understanding of the self-regulation of different cells is needed, and this should be accompanied by the necessary biochemical and biophysical experiments.

References

1. A. Szent-Györgyi, Nature **148**, 157 (1941).
2. A. Szent-Györgyi, Science **93**, 609 (1941).
3. A. L. Lehninger, "Biochemistry", Worth Publ. Inc., New York, 1975.
4. See for instance D. D. Eley, G. P. Parfitt, M. B. Perry, and D. H. Taytum, Trans. Far. Soc. **49**, 79 (1953); C. T. O'Konski and M. Shirai, Biopolymers **1**, 557 (1953); P. Pethig and A. Szent-Györgyi, Proc. Natl. Acad. Sci. USA **74**, 226 (1977).
5. M. G. Evans and J. Gergely, Biochim. Biophys. Acta **3**, 188 (1949); M. Suard, G. Berthier and B. Pullman, Biochim. Biophys. Acta **52**, 400 (1962); J. Ladik, Nature **202**, 1208 (1964).
6. L. Brillouin, in "Horizons in Biochemistry", M. Kasha, B. Pullman, eds., Interscience, New York, 1962.
7. A. Szent-Györgyi, "Introduction to a Submolecular Biology", Academic Press, New York, London, 1960; A. Szent-Györgyi, "Bioelectronics", Academic Press, New York, London, 1968; A. Szent-Györgyi, "Electronic Biology and Cancer", Marcel Dekker Inc., New York, Basel, 1976.
8. Y.-J. Ye and J. Ladik, Phys. Rev. **B48**, 5120 (1993).
9. Y.-J. Ye and J. Ladik, Intern. J. Quantum Chem. Proc. of the I. Congress of the International Society for Theoretical Chemical Physics, 1993 (in press).
10. P. Dean, Proc. Roy. Soc. **A254**, 507 (1960); ibid. **A260**, 263 (1961); Rev. Mod. Phys. **44**, 922 (1972); M. Seel, Chem. Phys. **43**, 103 (1979); J. Ladik, M. Seel, P. Otto, and A. K. Bakhshi, Chem. Phys. **108**, 203 (1986).
11. R. S. Day and F. Martino, Chem. Phys. Lett. **84**, 86 (1981).
12. Y.-J. Ye, J. Math. Chem. **14**, 121 (1993).

13. J. H. Wilkinson, "The Algebraic Eigenvalue Problem", Clarendon Press, Oxford, 1965, pp. 633.
14. A. K. Bakhshi, J. Ladik, M. Seel, and P. Otto, Chem. Phys. **108**, 233 (1986).
15. T. Odagaki and M. Lax, Phys. Rev. **B24**, 5284 (1981); ibid. **B26**, 6480 (1982).
16. N. F. Mott and E. A. "Electronic Processes in Non-Crystalline Materials", Clarendon Press, Oxford, 1971, p. 215.
17. O. Warburg, "Prevention of Cancer" in "Reflections on Biological Research", B. Gabbiani, ed., Warren H. Green Inc., St. Louis, 1967.

Index

Springer-Verlag and the Environment

We at Springer-Verlag firmly believe that an international science publisher has a special obligation to the environment, and our corporate policies consistently reflect this conviction.

We also expect our business partners – paper mills, printers, packaging manufacturers, etc. – to commit themselves to using environmentally friendly materials and production processes.

The paper in this book is made from low- or no-chlorine pulp and is acid free, in conformance with international standards for paper permanency.